Biostatistics and Computer-based Analysis

of Health Data using Stata

Biostatistics and Health Science Set

coordinated by
Mounir Mesbah

Biostatistics and Computer-based Analysis of Health Data using Stata

Christophe Lalanne
Mounir Mesbah

ELSEVIER

First published 2016 in Great Britain and the United States by ISTE Press Ltd and Elsevier Ltd

ISTE Press Ltd
27-37 St George's Road
London SW19 4EU
UK

www.iste.co.uk

Elsevier Ltd
The Boulevard, Langford Lane
Kidlington, Oxford, OX5 1GB
UK

www.elsevier.com

For information on all our publications visit our website at http://store.elsevier.com/

British Library Cataloguing-in-Publication Data
A CIP record for this book is available from the British Library
Library of Congress Cataloging in Publication Data
A catalog record for this book is available from the Library of Congress
ISBN 978-1-78548-142-0

Printed and bound in the UK and US

Contents

Introduction

A large number of the actions performed by means of statistical software are essentially forms of manipulating, or even literally transforming digital data representing statistical data. It is therefore paramount to fully understand how statistical data are represented and how they can be employed by software such as Stata. After the importing, recoding and the eventual transformation of these data, the description of the variables of interest and the summary of their distribution in numerical and graphical form constitute a fundamental preparatory stage to any statistical modeling, hence the importance of these early stages in the progress of a project for statistical analysis. In a second step, it is essential to fully control the commands that enable the calculation of the main measures of association in medical research, and to know how to implement the conventional explanatory and predictive models: analysis of variance, linear and logistic regression and the Cox model. With a few exceptions, making use of the Stata commands available during the installation of the software (base commands) will be preferred over the usage of specialized libraries of commands.

This book assumes that the reader is already familiar with basic statistical concepts, in particular the calculation of central tendency and dispersion indicators for a continuous variable, contingency tables, analysis of variance and conventional regression models. The objective here is to apply this knowledge to datasets described in numerous other works, even if the interpretation of the results remains minimal, in order to quickly familiarize oneself with the use of Stata with actual data. Emphasis is particularly given to the management and the manipulation of structured data since it can be noted that this constitutes 60–80% of the work of the statistician. There are many books in French or in English on Stata, covering both the technical and the statistical point of view. Some of these works show a dominant generalistic nature [ACO 14, HAM 13, RAB 04], while others are much more specialized and address similar topics, such as [FRY 14, DUP 09, VIT 05]. The purpose of this book is to enable the reader to quickly become accustomed to Stata, so that they can perform their own analyses and continue learning in an autonomous way in the field of medical statistics.

This book constitutes a sequel to the book *Biostatistics and Computer Analysis of Health Data using R* [LAL 16], published by the same authors in the same collection. Every topic that relates to data organization and data exploratory analysis, in particular graphical methods, are discussed therein. In this book, the same data sets are being used to facilitate the transfer of learning of the knowledge acquired in R.

In Chapter 1, the base commands for data management with Stata will be introduced. This primarily concerns the creation and the manipulation of quantitative and qualitative variables (recoding of individual values, counting of missing observations), importing databases stored in the form of text files, as well as elementary arithmetic operations (minimum, maximum, arithmetic mean, difference, frequency, etc.). We will also examine how to store preprocessed databases in text or in Stata formats. The objective is to understand how data are represented in Stata and how to work with them. The useful commands for describing a data table composed of quantitative or qualitative variables are also presented. The descriptive approach is strictly univariate, which constitutes the prerequisite for any statistical approach. Base graphic commands (histograms, density curves, bar or dot plots) will be presented in addition to the usual central tendency (mean, median) and dispersion (variance, quartiles) numerical descriptive summaries. Pointwise and interval estimation using arithmetic means and empirical proportions will also be addressed. The objective is to become familiar with the use of simple Stata commands operating on a variable, optionally specifying certain options for the calculation, alongside the selection of statistical units among all of the available observations.

Chapter 2 is dedicated to the comparison of two samples for quantitative or qualitative measurements. The following hypothesis tests are addressed: the Student's test for independent or paired samples, the non-parametric Wilcoxon test, the χ^2 test and the Fisher's exact test, and the NcNemar test based on the main measures of association for two variables (average difference, odds ratio and relative risk). From this chapter onwards, there will be less emphasis on the univariate description of each variable, but it is advisable to always carry out the stages of data description discussed in this chapter. The objective is to control the main statistical tests in the case where the relationship between a quantitative variable and a qualitative variable, or for two qualitative variables, is the main interest. This chapter also presents analysis of variance (ANOVA) where we explain the variability observed at the level of a numerical response variable by taking a group or classification factor into account, and the estimation with confidence intervals of average differences. Emphasis will be placed on the construction of an ANOVA table summarizing the various sources of variability, and on the graphic methods that can be used to summarize the distribution of individual or aggregated data. The linear tendency test will also be studied when the classification factor can be considered as naturally ordered. The objective is to understand how to construct an explanatory model in the case where there is one or even two explanatory factors, and how to digitally and graphically present the results of such a model through the use of Stata.

Chapter 3 focuses on the analysis of the linear relation between two continuous quantitative variables. In the linear correlation approach, which assumes a symmetrical relation between the two variables, the main focus will be on quantifying the force and the direction of the association in a parametric (Pearson correlation) or in a non-parametric manner (rank-based Spearman correlation) and on the graphic representation of this relation. Simple linear regression will be used in the event that one of the two numeric variables assumes the function of a response variable, and the other that of an explanatory variable. The useful commands for the estimation of the coefficients of the regression line, the construction of the ANOVA table associated with the regression and the computation of fitted values will be presented. The objective of this chapter remains identical to that of Chapter 2, namely to present the Stata commands necessary for the construction of a simple statistical model between two variables following an explanatory or predictive perspective.

In Chapter 4, the main measures of association found in epidemiological studies will be discussed: odds ratio, relative risk, prevalence, etc. Stata commands allowing the estimation (pointwise and by interval) and the associated hypothesis tests will be illustrated with data from cohort or case–control studies. The implementation of a simple logistic regression model makes it possible to complete the range of statistical methods, allowing the observed variability to be explained at the level of binary response variables. The objective is to understand the Stata commands to be used in the case in which the variables are binary, either to summarize a contingency table in the form of association indicators or to model the relationship between a binary response (ill/healthy) and a qualitative explanatory variable based on the so-called grouped data.

Chapter 5 constitutes an introduction to the analysis of censored data, the main tests associated with the construction of a survival curve (log-rank or Wilcoxon tests) and finally the Cox regression model. The specificity of the censored data requires particular care in the coding of data in Stata, and the objective is to present the Stata commands essential to the correct representation of survival data in digital form, to their numerical (survival median) and graphical (Kaplan–Meier curve) summary, and the implementation of common tests.

At the end of each chapter, a few applications are provided and a few examples of commands that can be used to respond to most of the presented questions are proposed. It is sometimes possible to obtain identical results with other approaches or by utilizing other commands. Stata outputs are not reproduced but readers are encouraged to try themselves the proposed Stata instructions and to try alternative or complementary instructions. It will be assumed that the data files used are available in the working directory. All of the data files and the Stata commands used in this book can be downloaded from the companion website (https://github.com/biostatsante).

Due to layout reasons, some of the Stata outputs have been truncated or reformatted. As a result, these could present differences when the reader attempts to reproduce the commands mentioned in this book.

An index of the Stata commands used in the illustrations is available at the end of the book.

Language Elements

In this chapter, the main topic will be the mode of representation of data in Stata and their manipulation. In particular, we will see how to represent numerical variables and categorical variables, how to operate on subsets of observations or how to only select parts that verify logical conditions, and finally the base syntax of Stata instructions (if, in, by).

1.1. Data representation in Stata

The data manipulated in Stata are mainly of two types: numbers and character strings. The numbers can be integers or real numbers. The first type is also used to encode the levels of a categorical variable to which text labels can be associated, called "variable labels" in Stata.

1.1.1. *The Stata language*

There are controls that allow users to easily generate a series of random numbers. The following example helps to familiarize with the basic elements of the Stata language. The following series of instructions allows storing in a variable called x 10 observations obtained from a normal distribution of average 12 and standard deviation 2:

```
. set obs 10
. generate x = rnormal(12, 2)
. format x %6.3f
. summarize x, format
obs was 0, now 10
```

```
Variable |        Obs        Mean    Std. Dev.         Min         Max
-------------+------------------------------------------------------------
        x |         10      11.112       2.246       7.956      14.224
```

Several remarkable features of the language should be noted: it is necessary to indicate the size of the sample used. In the following sections, we will see how these data can be obtained during manual input or when importing an external data file. The command generate can be used to associate with a variable, here x, a sequence of numeric values (assimilated here to our 10 observations) provided by the function (or subcommand) rnormal(). This latter has options available which enables the user to specify parameters of the distribution (mean and standard deviation, respectively) be specified. The command format x makes it possible to limit the display to 3 decimal places: this is a property of representation of the values of x directly associated with the variable that the command summarize can use.

Individual data can be examined by means of the command list. For example, the command list x will display all of the values of x. Since there is only a single variable present in the Stata workspace, it is nonetheless equivalent to typing list for short. The option in can be included to restrict the display of the values of x to the fifth observation or to the first five observations. In the latter case, the ranks of the observations are indicated in the first value/last value form: the expression 1/5 therefore designates the observations numbered 1–5:

```
. list x in 5
     +-------+
     |   x   |
     |-------|
  5. | 7.956 |
     +-------+
. list x in 1/5
     +--------+
     |    x   |
     |--------|
  1. | 11.118 |
  2. | 13.889 |
  3. | 14.224 |
  4. |  8.726 |
  5. |  7.956 |
     +--------+
```

1.1.2. *Creating and manipulating variables*

In the case of small datasets, it is possible for users to enter themselves the observations, although most of the time it will be preferable to work from an external file. For this purpose, the command input is available which is employed in the following manner: after the name of the command, the name of the variable(s) is indicated, separated by a space, and then the user ought to press the Enter key before entering the data, always separated by spaces. To indicate to Stata that the entry is complete, the word end must be inserted. This manual entry can also be performed from the data editor (Data ▷ Data Editor ▷ Data Editor (Edit)). Here is an example of the usage with a series of 10 weight measurements collected in newborns (x, in grams) and their mother (y, in kilograms).

x	2523	2551	2557	2594	2600	2622	2637	2637	2663	2665
y	82.7	70.5	47.7	49.1	48.6	56.4	53.6	46.8	55.9	51.4

Table 1.1. *Artificial data on the weight at birth*

Data input in Stata would thus be achieved as follows: enter input x y in the Stata console, and then for the line numbered 1. indicate 2523 82.7 and type Enter, followed by the line numbered 2. indicate 2551 82.7 and Enter and so on until the 10th line. For the 11th line, we will simply write end and press Enter. The result should look like the following:

```
. input x y

         x    y
 1. 2523 82.7
 2. 2551 70.5
 3. 2557 47.7
 4. 2594 49.1
 5. 2600 48.6
 6. 2622 56.4
 7. 2637 53.6
 8. 2637 53.6
 9. 2637 46.8
10. 2663 55.9
11. 2665 51.4
12. end
```

It is possible to transform the values taken by a variable or to create new variables based on the values taken by a variable by using the commands generate and replace. This last command works exclusively on an existing variable. Here is an example of using generate where the weight of infants (x) is converted into kilograms:

```
. generate x2 = x / 1000
. list x x2 in 1/3
```

```
     +---------------+
     |   x      x2  |
     |---------------|
  1. | 2523    2.523 |
  2. | 2551    2.551 |
  3. | 2557    2.557 |
     +---------------+
```

It is also possible to replace the set of values from any transformation, for example the logarithm:

```
. replace x 2 = log (x
2) (10 real changes made)
```

or specifically override certain values indicating an observation number, as illustrated hereafter:

```
. replace x2 = 2600 in 3
(1 real change made)
. list x x2 in 1/3
```

```
     +-------------------+
     |   x        x2    |
     |-------------------|
  1. | 2523    .9254487 |
  2. | 2551    .9364855 |
  3. | 2557       2600 |
     +-------------------+
```

Finally, the command drop can be inserted to delete any variable of the workspace. In this case, the observations are then permanently lost:

```
. drop x2
```

1.1.3. *Indexed or criteria-based selection of observations*

The option for the selection of observations based on indices (or ranks) has already been presented with the option in:

```
. list x in 1/3
     +------+
     |   x  |
     |------|
  1. | 2523 |
  2. | 2551 |
  3. | 2557 |
     +------+
```

It is quite possible to select observations on the basis of an external criterion, for example the values taken by a second numerical variable. In the following example, only the weight of babies for which the mothers' weight ≤ 50 is retained:

```
. list x if y <= 50
     +------+
     |   x  |
     |------|
  3. | 2557 |
  4. | 2594 |
  5. | 2600 |
  8. | 2637 |
     +------+
```

Suppose that there is also some information available concerning the fact that the mother was smoking during the first trimester of pregnancy; we will call this variable z. When the mother did not smoke during this period, the variable is equal to 1; when the mother was smoking, the variable is 2. In the following, the previous data table is presented amended to account for this information.

x	2523	2551	2557	2594	2600	2622	2637	2637	2663	2665
y	82.7	70.5	47.7	49.1	48.6	56.4	53.6	46.8	55.9	51.4
z	1	1	2	2	2	1	1	1	2	2

Table 1.2. *Augmented artificial data about the weights at birth*

The new data entry does not raise any difficulty and once again input will be used, indicating z as a new variable. The end of the input should be signaled by means of the instruction end (followed by Enter). Here follows an overview of the first five observations for the three variables:

```
. list in 1/5
        +-----------------+
        |   x     y    z  |
        |-----------------|
    1.  | 2523   82.7   1 |
    2.  | 2551   70.5   1 |
    3.  | 2557   47.7   2 |
    4.  | 2594   49.1   2 |
    5.  | 2600   48.6   2 |
        +-----------------+
```

It is possible to refine our search criteria by restricting the selection of observations x depending on the values taken by y and z. The following instruction displays the weight of babies whose mother weighs less (strictly) than 55 kg and who was not smoking during her pregnancy:

```
. list x if y < 55 & z == 1
        +------+
        |  x   |
        |------|
    7.  | 2637 |
    8.  | 2637 |
        +------+
```

It can be seen that there are two statistical units that verify the previous conditions ($y < 55$ and $z = 1$). The logical "and" (conjunction) is denoted &, while the logical "or" (disjunction) is written |. In order to count the observations verifying the preceding condition, it is possible to employ the command count:

```
. count if y < 55 & z == 1
    2
```

1.1.4. *Processing the missing values*

Missing values are represented by a dot in Stata. For example, it is possible to replace the third observation of x with a missing value, by using the command replace presented above:

```
. replace x = . in 3
```

```
(1 real change made, 1 to missing)
. summarize x

    Variable |      Obs        Mean    Std. Dev.       Min        Max
-------------+---------------------------------------------------------
         x |        9    2610.222    48.53035       2523       2665
```

It shall be verified that the number of observations reported by summarize is correctly accounting for the missing datum ($n = 9$ instead of 10). We can also verify the number of missing values identified for a variable by using the command misstable:

```
. misstable summarize x

                                                     Obs<.
                                          +----------------------------
             |                            | Unique
    Variable |    Obs=.      Obs>.   Obs<. | values       Min        Max
-------------+----------------------------+----------------------------
         x |        1               9 |     8        2523       2665
-----------------------------------------------------------------------
```

In practice, the missing data are represented in the form of a very large number; care should be taken when performing numerical comparison tests, and when in doubt always use a test of the type if !missing(x) (more elegant than if x< .) in order to be certain that you are working with the observed data:

```
. summarize y if !missing(x)
    Variable |      Obs        Mean    Std. Dev.       Min        Max
-------------+---------------------------------------------------------
         y |        9    57.22222    11.85219       46.8       82.7
```

1.1.5. *Data management*

1.1.6. *Importing external data*

There are several commands that enable importing the data contained in a text file. For files in which the fields are separated by one or more spaces, we will make use of the command infile. In the case where there is a field separator such as a comma (typical of CSV files exported from an Excel spreadsheet) or a tabulation, the command insheet will be employed, eventually specifying the type of the field delimiter with the option delimiter(). The command insheet works well in the case where the first line of the file contains the name of the variables. However, in both cases, it is possible to provide a list for the names of the variables, and the name of the file will always be indicated after the instruction using. The representation format of the data that was read can also be customized by specifying its type before each variable, for example, with the command:

```
. infile str5 name age byte rep using "fichier.txt", clear
```

Stata is instructed to build from the file called fichier.txt a table containing three variables: name, age and rep. The variable name must be explicitly treated as a string of characters (maximum five) and the storage format of the variable rep must be limited to the minimum (1 byte = values varying from -127 to -100), for instance so as not to unnecessarily occupy memory. Finally, in some cases, we may avoid inserting options at the command line (variables name and storage format) and store all of this information in what is called a dictionary, see help infile2. Note that Stata provides dialog boxes that make it possible to specify different options for each of these commands, such as the field delimiter or the presence of a header row, and that a data previewer allows verifying whether the data structure has been properly defined.

Consider data on birth weight, available in a file called birthwt.dat in which fields are separated by a space as in the following overview:

```
0 19 182 2 0 0 0 1 0 2523
0 33 155 3 0 0 0 0 3 2551
0 20 105 1 1 0 0 0 1 2557
0 21 108 1 1 0 0 1 2 2594
0 18 107 1 1 0 0 1 0 2600
```

It should be noted that the name of the variables does not appear on the first line of the file. Each column corresponds, respectively, to the following variables: weight status of the baby at birth low (= 1 if weight < 2.5 kg, 0 otherwise), age of the mother (years), lwt weight of the mother (in pounds), race ethnicity of the mother (encoded in three classes, 1 = white, 2 = black, 3 = other), smoke (= 1 if tobacco consumption during pregnancy, 0 otherwise), ptl (number of previous preterm births), ht (= 1 if hypertension history, 0 otherwise), ui (= 1 if intrauterine irritability, 0 otherwise), ftv (number of consultations with the gynecologist during the first pregnancy trimester), bwt for the weight of babies at birth.

The command infile can be used to import these data, indicating the list of variables that Stata will associate to each column of the external data file. The option clear instructs Stata to delete the existing data in the workspace before importing:

```
. infile low age lwt race smoke ptl ht ui ftv bwt using "birthwt.dat", clear
(189 observations read)
. list in 1/5
```

```
+------------------------------------------------------------+
| low   age   lwt   race   smoke   ptl   ht   ui   ftv   bwt |
|------------------------------------------------------------|
1. |  0    19   182     2       0     0    0    1     0   2523 |
2. |  0    33   155     3       0     0    0    0     3   2551 |
3. |  0    20   105     1       1     0    0    0     1   2557 |
4. |  0    21   108     1       1     0    0    1     2   2594 |
5. |  0    18   107     1       1     0    0    1     0   2600 |
+------------------------------------------------------------+
```

In order to display the observations that were read and that are now contained in the workspace, we will use the command list. As previously seen, it is possible to limit the number of displayed observations ("rows") by including a filter on the numbers of observations: the option in 1/5 instructs Stata to select only the observations ranging from 1 to 5. It is also possible to open the data viewer in order to obtain a view similar to an Excel table for these data.

1.1.7. *Variable management*

The data imported into the workspace can also be displayed by inserting the command describe. With the option simple, Stata returns the name of the variables only, whereas with the option short we get a summary indicating the number of observations and variables:

```
. describe, short
Contains data
  obs:           189
  vars:           10
  size:         7,560
Sorted by:
```

It is also possible to provide a list of variables, for example the variables low, age and lwt. These are the first three variables and the expression describe low age lwt can then be simplified into describe low-lwt:

```
. describe low-lwt
                storage   display     value
variable name   type      format      label      variable label
-------------------------------------------------------------------
low             float     %9.0g
age             float     %9.0g
lwt             float     %9.0g
```

It can be seen that these three variables (weight indicator $< 2,500$ g, mother's age and mother's weight, in pounds) are manipulated as numbers. Due to economy concerns regarding memory space, the following command might be preferred:

```
. infile byte low age lwt race smoke ptl ht ui ftv bwt using "birthwt.dat", clear
(189 observations read)
```

to inform Stata to reserve less memory space for the variable low, which is a binary variable (the name of the variable has been prefixed by the statement byte).

To associate labels with variables, we will use the command label variable:

```
. label variable low "Weight smaller than 2.5 kg"
. describe low-lwt race

              storage  display    value
variable name  type   format     label      variable label
--------------------------------------------------------------------------

low            byte   %8.0g                 Weigh less than 2.5 kg
age            float  %9.0g
lwt            float  %9.0g
race           float  %9.0g
```

The variable race, although categorical, is represented as a number by Stata (float), and it is possible to verify its values with the command tabulate that provides a frequency table:

```
. tabulate race
     race |      Freq.     Percent       Cum.
----------+-----------------------------------
        1 |         96       50.79      50.79
        2 |         26       13.76      64.55
        3 |         67       35.45     100.00
----------+-----------------------------------
    Total |        189      100.00
```

The modalities or levels of the categorical variables assumed as numbers can be associated with labels, which facilitates the reading of the descriptive graphs and summary tables. To this end, it is necessary to define, as a first step, the correspondence between the values of the variable and the labels (label define), then, in a second step, to associate these labels to the variables (label values). Here follows how to proceed with the variables race, ht and ui:

```
. label define yesno 0 "No" 1 "Yes"
```

```
. label define ethn 1 "White" 2 "Black" 3 "Other"
. label values ht ui yesno
. label values race ethn
```

The usage of label define is rather straightforward: the name of the label that will serve as a reference is given and each value is associated with a description in the form of characters (0 is here associated with ''No''). Similarly, for label values, the variables are indicated followed by the reference created with label define:

```
. tabulate race
      race |      Freq.     Percent        Cum.
-----------+-----------------------------------
     White |         96       50.79       50.79
     Black |         26       13.76       64.55
     Other |         67       35.45      100.00
-----------+-----------------------------------
     Total |        189      100.00
```

1.1.8. *Converting a numerical variable into a categorical variable*

When the bounds of the class intervals that must be considered are known, the command egen cut can be utilized, indicating the lower bounds of the intervals. Here follows a usage example:

```
. egen lwt3 = cut(lwt), at(70,120,170,220,270)
. tabulate lwt3
      lwt3 |      Freq.     Percent        Cum.
-----------+-----------------------------------
        70 |         75       39.68       39.68
       120 |         93       49.21       88.89
       170 |         17        8.99       97.88
       220 |          4        2.12      100.00
-----------+-----------------------------------
     Total |        189      100.00
```

The command egen constitutes an extension of the command generate allowing the creation of new variables, but accepting a certain number of options (acting most of the time as functions that enable performing calculations on a given variable). See online help, help egen, for more information and in particular the list of functions available (count, iqr, max, etc.).

Although all of the bounds of the intervals have been explicitly specified, it would also be possible to write at(70(50)270) to inform Stata to build a sequence of values ranging from 70 to 270 with increments of 50. Stata can also automatically build more or less balanced groups with the option group(4).

In order to build classes based on quartiles or on deciles, the command xtile should be employed instead, specifying in the option nq() the number of desired groups:

```
. drop lwt3
. xtile lwt3 = lwt, nq(4)
. tabulate lwt3
4 quantiles |
     of lwt |       Freq.     Percent         Cum.
------------+-----------------------------------
         1 |          53       28.04        28.04
         2 |          43       22.75        50.79
         3 |          46       24.34        75.13
         4 |          47       24.87       100.00
------------+-----------------------------------
     Total |         189      100.00
```

1.2. Descriptive univariate statistics and estimation

1.2.1. *Summarizing a numerical variable*

The command summarize provides a four-point numerical summary (average, standard deviation, minimum and maximum) for one or more numerical variables. The option detail provides a more comprehensive summary, notably including the five most extreme values, different quantiles and the indicators of skewness and kurtosis indicators of the distribution of the variable:

```
. summarize bwt
    Variable |        Obs        Mean    Std. Dev.        Min         Max
-------------+-------------------------------------------------------------
         bwt |        189    2944.587    729.2143         709        4990
```

To obtain a 95% confidence interval based on the approximation by the normal distribution, we will insert the command ci followed by the name of the variable of interest:

```
. ci bwt
    Variable |        Obs        Mean    Std. Err.      [95% Conf. Interval]
-------------+-------------------------------------------------------------
         bwt |        189    2944.587    53.04254       2839.952    3049.222
```

The command `mean bwt` will provide the same result.

To build a histogram of frequencies or frequency counts, the command to employ is `histogram`. The option `frequency` should be included when willing to work with frequency counts, or `percent` for proportions (Figure 1.1):

```
. histogram bwt, frequency
(bin=13, start=709, width=329.30769)
```

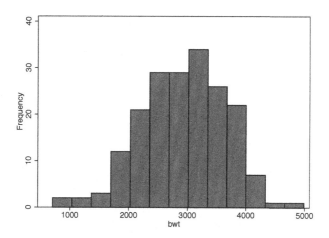

Figure 1.1. *Frequency counts histogram for the weight at birth*

The option `bin()` make it possible to change the number of class intervals used by Stata. When willing to display a nonparametric density curve, we will add the option `kdensity`. There is also an option `discrete` that assimilates the histogram to a representation in the form of a bar plot in which are represented on the x-axis, the unique values of the numerical value, without consideration of class intervals.

1.2.2. *Summarizing a categorical variable*

The command `tabulate` (which can be simplified in `tab`) provides a frequency analysis table for a variable:

```
. tabulate race, plot
       race |      Freq.
------------+------------+----------------------------------------------------
      White |        96 |***************************************************
      Black |        26 |*************
```

```
    Other |        67 |***********************************
------------+------------+-----------------------------------------------------
    Total |       189
```

tab1 is preferable when we want to build (univariate) frequency counts tables for several variables, with the following syntax:

```
. tab1 race ht ui
```

As in the case of numeric variables, the command ci (or prop) provides confidence intervals for a proportion. The option binomial will be added if it is desirable to use the binomial distribution to build 95% confidence intervals:

```
. ci low, binomial

                                            -- Binomial Exact --
    Variable |      Obs      Mean    Std. Err.    [95% Conf. Interval]
-------------+--------------------------------------------------------
        low |      189   .3121693   .0337058     .2468886    .3834546
```

The option level(), as in most of the commands related to statistical tests of null hypothesis or to interval-based estimation, allows changing the degree of significance associated with the confidence intervals. For example, level(90) instructs Stata to return 90% confidence intervals, instead of 95%, which is the value by default.

There exist several ways to graphically represent frequency counts or frequency distributions. Hereafter follows a solution for the variable race that is based on the use of the command graph bar (Figure 1.2). The idea consists of creating an auxiliary variable in which the counts associated with each category of race are accumulated:

```
. gen freq = 1
. graph bar (sum) freq, over(race) ytitle("Ethnicity")
```

The instruction bar will be replaced by hbar to represent the bars horizontally instead of vertically. Another solution is to use the command histogram with the option discrete.

1.3. Bivariate descriptive statistics

1.3.1. *Describing a numeric variable by the levels of a categorical variable*

The command summarize operates in a univariate manner only, that is for each listed variable. When a numeric variable has to be summarized for each level of a categorical variable, a selection option by can be used, which has to be placed at the beginning of the command:

```
. by low, sort: summarize lwt
-------------------------------------------------------------------------
-> low = 0

    Variable |       Obs        Mean    Std. Dev.       Min        Max
-------------+-----------------------------------------------------------
         lwt |       130       133.3    31.72402        85         250


-------------------------------------------------------------------------
-> low = 1

    Variable |       Obs        Mean    Std. Dev.       Min        Max
-------------+-----------------------------------------------------------
         lwt |        59    122.1356    26.55928        80         200
```

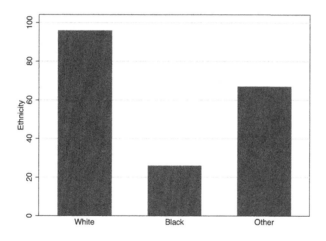

Figure 1.2. *Bar diagram for the distribution of counts of the variable* race

It should be noted that the data must be sorted first, hence the addition of the option sort. An alternative consists of directly using bysort:

```
. bysort low: summarize
```

Care should be taken not to confuse: with , when by is specified before a command.

It may happen that only certain statistics have to be calculated, for example the mean and the standard deviation. In this case, the command `tabstat` is simpler to use. Here is an example of its usage:

```
. tabstat lwt, by(low) stats(mean sd) format(%6.2f)
Summary for variables: lwt
      by categories of: low (Weight less than 2.5 kg)

    low |      mean         sd
---------+--------------------
      0 |    133.30      31.72
      1 |    122.14      26.56
---------+--------------------
  Total |    129.81      30.58
----------------------------
```

The option `format(%6.2f)` makes it possible to limit the display to two decimal places.

An alternative formulation which can be generalized to several classification factors (potentially using different statistics, for example the average for a variable and the median for another), consists of using the command `table`. The equivalent to the option `stats()` of `tabstat` is here `contents()` and what is to be calculated is specified therein: `freq` corresponds to the count by the level of the classification variable `low`, `mean lwt` corresponds to the average of the variable `lwt` for each level of the classification factor, etc.:

```
. table low, contents(freq mean lwt sd lwt) format(%6.2f)
------------------------------------------------
Weight     |
less       |
than 2.5   |
kg         |    Freq.    mean(lwt)    sd(lwt)
---------+--------------------------------------
      0 |   130.00      133.30       31.72
      1 |    59.00      122.14       26.56
------------------------------------------------
```

In order to represent the average weight of mothers for underweight children or within standards, a dot plot (or even a bar plot, as in the previous case with the variable `race`) is appropriate: the classification factor is indicated in an option `over()`, which allows overlaying graphical elements within the same graph as illustrated in Figure 1.3:

```
. graph dot lwt, over(low)
```

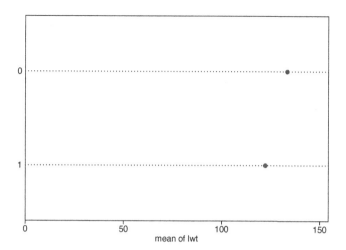

Figure 1.3. *Dot chart for the average values of*
lwt conditionally to the variable low

By default, it is the average of the numeric variable (`lwt`) that is considered, but another statistic can be used by inserting the option (`stats`). For example, when the median weight must be displayed, the previous command will be replaced by:

```
. graph dot (median) lwt, over(low)
```

In some cases, it may happen that only a particular summary statistic is of interest. For example, suppose that we want to calculate the maximal weight of the babies (in grams) in the two groups of individuals defined by the variable `ui` (0, no intrauterine pain during pregnancy; 1, intrauterine pain). In this case, it is possible to store these two values in variables known as "local", by making use of the instruction `scalar`. In order to calculate and display the maximum weight observed in babies in the group of mothers presenting no intrauterine pain, the following commands should be used:

```
. drop maxbwt
. quietly: summarize bwt if ui == 0
. scalar maxbwt0 = r(max)
. display maxbwt0
4990
```

In fact, we use the results generated by the command `summarize` but recorded temporarily in an invisible manner by Stata (they can be accessed by typing `return list`, just after typing `summarize bwt if ui == 0`). The content of interest is retrieved by inserting the command `r()` and it is then possible to display the results with `display`. The addition of the prefix `quietly:` to the command `summarize` ensures that the results are not displayed (but that they still remain available).

1.3.2. *Describing two qualitative variables*

The command `tabulate` makes it possible to build contingency tables when it is invoked with a list of two variables. For example, the following command enables crossing the levels of the variables `low` (row-wise) and `smoke` (column-wise) and presents the relative frequencies calculated row-wise:

```
. tabulate low smoke, row
+-----------------+
| Key             |
|-----------------|
|   frequency     |
| row percentage  |
+-----------------+

  Weight   |
 less than |       smoke
   2.5 kg  |       0          1 |      Total
-----------+----------------------+----------
        0  |      86         44 |        130
           |   66.15      33.85 |     100.00
-----------+----------------------+----------
        1  |      29         30 |         59
           |   49.15      50.85 |     100.00
-----------+----------------------+----------
    Total  |     115         74 |        189
           |   60.85      39.15 |     100.00
```

The other options, `col` and `cell`, allow the calculation of the relative frequencies in a columnwise manner or relatively to the total number of the observations (conditional frequencies).

1.4. Key points

– Stata represents data as a list of variables, similar to a data table in which the variables are arranged in columns, all having the same number of observations, and the values of the variables are generally numbers that can be associated with labels when they refer to the modalities of a categorical variable.

– The commands summarize and tabulate provide a univariate descriptive summary in the case of numeric and categorical variables, whereas tabstat and tabulate allow working in a bivariate manner.

– The main graphics commands to represent the distribution of a numeric or categorical variable are histogram and graph bar (bar plot) or graph dot (dot plot).

1.5. Further reading

Stata is a well-known software program to facilitate database management. For more information, the manual "[D] data management", accessible in PDF format or by using the command help data management, provides an accurate description of all the controls related to the management and manipulation of variables.

With regard to the production of graphs in Stata, it is strongly recommended to consult the work of Mitchell [MIT 12], which provides a detailed description of each type of graph with the most common options. The Stata website also offers an overview of the different charts available depending on the type of variables manipulated: http://www.stata.com/support/faqs/graphics/gph/stata-graphs/.

Finally, for all aspects related to automation or programming with Stata, the reader is invited to refer to the book by Baum [BAU 16].

1.6. Applications

1) The plasma viral load is used to describe the amount of virus (for example HIV) in a blood sample. This viral marker that allows following the progression of the infection and measuring the effectiveness of treatments represents the number of copies per milliliter, and most measurement instruments have a detectability threshold of 50 copies/mL. Here follows a series of measurements, X, expressed in logarithms (base 10) collected on 20 patients:

```
3.64 2.27 1.43 1.77 4.62 3.04 1.01 2.14 3.02 5.62 5.51 5.51 1.01
1.05 4.19 2.63 4.34 4.85 4.02 5.92
```

As a reminder, a viral load of 100,000 copies/mL is equivalent to 5 log.

– indicate how many patients have a viral load considered as non-detectable;

– the researcher realizes that the value 3.04 corresponds to a data entry error and must be changed to 3.64. Similarly, she has a doubt about the seventh measurement and decides to consider it as a missing value: perform the corresponding transformations;

– what is the median viral load level in copies per milliliter, for the data considered as valid?

Stata provides a data editor that comes in the form of a spreadsheet program (similar to Excel), but data can be directly recorded by utilizing the command input. After having given the name of the variables (when there are several variables, their name has to be separated with a space), the user presses the Enter button and the observations are input (similarly, when there are several variables the observations or observed values are typed in separated by a space). When there are no more data to be entered, the user can type the word end and press Enter to inform Stata that data input is completed. It is also possible to use the data editor, but in this case the user will have to rename the variable (by default var1).

The detection threshold in logarithm is:

```
. display log10(50)
```

It is therefore possible to count the number of observations that do not verify the condition $X > \log(50)$ by employing the command count:

```
. count if X <= log10(50)
```

Hence, the calculation of the median viral load considering only the data above the detection threshold:

```
. egen Xm = median(10^X) if X > log10(50)
. display round(Xm)
```

To calculate the median of X verifying the condition $X > \log(50)$, the command egen has been directly used that provides a certain number of transformations and basic computing functions (see the online help, help egen).

2) The file anorexia.dat contains data from a clinical study in anorexic patients who have undergone one of the three following therapies: behavioral therapy, family therapy and control therapy [HAN 93].

– how many patients are there in total? How many patients are there per treatment group?

– the weight measures are in pounds. Convert them into kilograms;

– create a new variable containing differences scores (After - Before);

– indicate the mean and the range (min/max) of the difference scores per treatment group.

The data contained in the file `anorexia.data` is read in the form of the overview provided below:

```
Group Before After
g1 80.5  82.2
g1 84.9  85.6
g1 81.5  81.4
g1 82.6  81.9
g1 79.9  76.4
```

we will provide the command `infile` with a "dictionary" file that describes the structure of the data. This file usually has the same name as the data source file, and has the extension `dct`. Here is the complete listing (`anorexia.dct`):

```
infile dictionary using anorexia.dat {
  _first(2)
  str2 Group "Therapy type"
  double Before "Before"
  double After "After"
}
```

Therein, we indicate that the observations begin on the second row (ignoring the header row), and that there are three variables, `Group` (qualitative variable), `Before` and `After` (numerical variables). Note that description labels have been associated for these three variables. It then suffices to just make use of the command `infile` by providing the name of the dictionary file (there is no need to specify the file extension):

```
. infile using anorexia
```

We can then verify that the data have been correctly imported by entering `describe`. This command also provides the number of observations available in the data table ($N = 72$):

```
. describe
```

To obtain the frequency counts by therapy type, a simple table of counts is produced by inserting the command `tabulate`:

```
. tabulate Group
```

The conversion of units for the weights does not raise any specific problem, but a decision must be made whether new variables have to be created (`generate`) or if the existing values are to be replaced (`replace`). Here, the existing values will be replaced:

```
. replace Before = Before/2.2
. replace After = After/2.2
```

For differences scores, this time a new variable is created by making use of the command generate:

```
. generate diff = After - Before
. summarize diff
```

The command summarize can be utilized to provide a numeric summary for each of the treatment groups, for example by Group, sort: summarize diff. However, to specifically calculate some descriptive indicators, it is more convenient to employ the command tabstat to which the response variable and the classification factor are provided, as well as the desired statistics through stats():

```
. tabstat diff, by(Group) stats(mean min max)
```

3) A quantitative variable X takes the following values with a sample of 26 individuals:

```
24.9,25.0,25.0,25.1,25.2,25.2,25.3,25.3,25.3,25.4,25.4,25.4,25.4,
25.5,25.5,25.5,25.5,25.6,25.6,25.6,25.7,25.7,25.8,25.8,25.9,26.0
```

– calculate the mean, the median and the mode of X;

– what is the value of the variance estimated from these data?

– assuming that data are grouped into four classes whose bounds are: 24.9–25.1, 25.2–25.4, 25.5–25.7, 25.8–26.0, display the distribution of the figures by class in the form of a frequency counts table;

– represent the distribution of X as a histogram, without consideration of *a priori* class intervals.

Concerning data entry, we can simply enter the data in a text format file. Suppose that the data have been entered in a file called saisie_x.txt of which an overview follows here:

```
24.9 25.0 25.0 25.1 25.2 ...
```

The values of X are simply separated by a space. In this case, the data can be read and imported into Stata with the command infile:

```
. infile x using "saisie_x.txt"
```

The option clear has to be added if data are already stored in the Stata workspace.

The command tabstat can be used to calculate the average and the median of the observations. However, the command egen must be included with the function mode to calculate the modal values of X. Since there are several modes, Stata will be instructed to return the lowest value of the two modes. Note that the command egen could also be employed to calculate the average and the median of X:

```
. tabstat x, stats(mean median)
. egen xmode = mode(x), minmode
. display xmode
```

Concerning the variance, summarize x provides direct access to the standard deviation, but it can also be calculated with egen as the square of the standard deviation estimated from the sample, and then displayed in the console of the results:

```
. egen varx = sd(x)
. di varx^2
```

A more elegant alternative consists of exploiting the results returned by a statement such as summarize x: in this case, display r(Var) provides the expected result.

Regarding the discretization of x into four predefined class intervals, the function cut can always be used along with egen. It can then be verified that the class intervals are properly respected by employing table, which is more flexible than the command summarize and makes it possible to specify the list of statistics values to be displayed:

```
. egen xc = cut(x), at(24.8,25.2,25.5,25.8,26.1) label
. table xc, contents(min x max x)
```

On the other hand, the frequency counts table can be directly obtained with tabulate, for example:

```
. tabulate xc, plot
```

Finally, Stata relies on a specific algorithm to determine the optimal number of classes to use in an histogram, just like R. Everything is managed from the command histogram:

```
. histogram x, frequency
```

The option frequency should not be forgotten when intending to display the counts instead of the density (which is the default choice).

4) The file elderly.dat contains the size, measured in centimeters, of 351 elderly females, randomly selected from the population during a study on osteoporosis. A few observations are however missing.

– how many missing observations are there in total?

– give a 95% confidence interval for the average size in this sample, using a normal approximation;

– represent the distribution of the sizes observed in the form of a density curve.

To import the data stored in a simple text file, the command `infile` is employed:

```
. infile tailles using "elderly.dat", clear
```

Here, it should be noted that the manner in which the missing values are coded is not specified because the "." is the default format by Stata (for numeric variables only).

There are several commands that facilitate the detection and the identification of the missing values. Among those available by default in Stata, `codebook` can be distinguished, that provides a summary of the variable, as well as the basic functions that allow the tabulation or counting the observations that meet a certain criteria:

```
. count if sizes == .
```

To obtain the average size and its associated 95% confidence interval, the following command has to be entered:

```
. ci sizes
```

Finally, to display the distribution of the sizes in the form of a density curve, the command `histogram` is inserted with the option `kdensity`. The degree of smoothing can be controlled with the option `kdenopts`; for example, adding `kdenopts(gauss width(1))` to the following command would produce a "less smooth" curve (that is to say, fitting better to the data):

```
. histogram sizes, kdensity
```

Measures of Association, Comparisons of Means and Proportions for Two Samples or More

This chapter focuses on measures of association between two categorical variables (χ^2 or Fisher test for a contingency table, and the calculation of the odds ratio (OR)) or between a numeric variable and a classification factor. In the latter case, we will consider the case of two independent (or not) samples, as well as parametric (Students t-test) and non-parametric (Wilcoxon test) models for two or more samples situations (analysis of variance (ANOVA) and Kruskal–Wallis ANOVA). The Bonferroni correction method for multiple comparisons of treatment and the linear trend test for the ANOVA will also be discussed. The case of two-factor ANOVA is presented succinctly, restricted to the major commands allowing for the construction of the ANOVA table and an interaction graph to be plotted.

2.1. Comparisons of two group means

2.1.1. *Independent samples*

The command `ttest` allows that a Student's t-test be performed for the comparison of two group means, assuming equal variances (or not) in the population (option `unequal`). When considering the mothers' weights on the basis of the indicator of babies born underweight, a brief descriptive summary (mean and standard deviation) can be obtained by means of the command `tabstat`, or as follows:

```
. format lwt %4.1f
. tabulate low, summarize(lwt)

            | Summary of weight at last menstrual
birthweight |                period
     <2500g |      Mean    Std. Dev.       Freq.
------------+-----------------------------------
          0 |     133.3        31.7         130
          1 |     122.2        26.5          59
------------+-----------------------------------
      Total |     129.8        30.6         189
```

In terms of graphical representation, the distributions of these two measurement series can be viewed through histograms:

```
. histogram lwt, by(low)
```

or box plots (Figure 2.1):

```
. graph box lwt, over(low, relabel(1 "Normal" 2 "Low (< 2.5 kg)")) ///
b1title("Baby weight") ytitle("Mother weight")
```

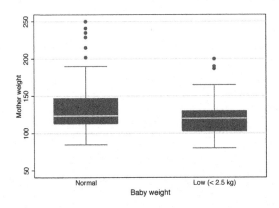

Figure 2.1. *Distribution of the mothers' weight in the form of box and whisker charts*

It should be noted that in the case of box plots (command graph box), there are two options available by() (displaying conditional distributions in juxtaposed graphics) and over() (displaying the conditional distributions in the same graph). The latter is not available for histogram.

The "conventional" Student's *t*-test is performed by specifying the response variable and the classification factor in an option by():

```
. ttest lwt, by(low)

Two-sample t test with equal variances
--------------------------------------------------------------------------------
   Group |     Obs        Mean    Std. Err.   Std. Dev.   [95% Conf. Interval]
---------+----------------------------------------------------------------------
       0 |     130       133.3     2.78238    31.72402     127.795     138.805
       1 |      59    122.1356    3.457723    26.55928    115.2142     129.057
---------+----------------------------------------------------------------------
combined |     189    129.8148    2.224323    30.57938     125.427    134.2027
---------+----------------------------------------------------------------------
    diff |            11.16441    4.743297                1.807157    20.52166
--------------------------------------------------------------------------------
    diff = mean(0) - mean(1)                                  t =    2.3537
Ho: diff = 0                                     degrees of freedom =       187

   Ha: diff < 0                 Ha: diff != 0                 Ha: diff > 0
Pr(T < t) = 0.9902       Pr(|T| > |t|) = 0.0196       Pr(T > t) = 0.0098
```

In the above approach, there are two well-defined variables available, one functioning as a response variable and the other as a classification factor. It is also possible to work with two series of measurements (not necessarily of the same size). Here is a possible approach, which serves mainly to demonstrate how a second data table can be managed without erasing the data present in the workspace by using the command `preserve` and `restore`. As a first step, two new variables will be created, lwt1 and lwt2, in which the weights of the two groups of mothers are stored:

```
. preserve
. gen lwt1 = lwt if low == 0
(59 missing values generated)
. gen lwt2 = lwt if low == 1
(130 missing values generated)
```

It is possible to verify the characteristics of these two variables, or even to compare them with all the data of the sample (lwt) using `summarize`: the notation lwt* informs Stata to consider all variables whose names begin with lwt (thus lwt, lwt1 and lwt2 in the present case):

```
. summarize lwt*

    Variable |     Obs        Mean    Std. Dev.       Min        Max
-------------+-----------------------------------------------------------
         lwt |     189    129.8148    30.57938         80        250
```

lwt1	130	133.3	31.72402	85	250
lwt2	59	122.1356	26.55928	80	200

The alternative formulation for the *t*-test is then:

```
. ttest lwt1 == lwt2, unpaired
```

In this case, it is important to specify the option unpaired. The command restore should not be forgotten in order to be able to return to the original data table (and delete the variables generated in the meantime).

The option welch provides a modified Welch *t*-test (whereas unequal is based on the Satterthwaite approximation). When we intend to formally test the hypothesis of equal variances by means of an *F*-test, the command sdtest can be used:

```
. sdtest lwt, by(low)
```

Variance ratio test

Group	Obs	Mean	Std. Err.	Std. Dev.	[95% Conf. Interval]
0	130	133.3	2.78238	31.72402	127.795 138.805
1	59	122.1356	3.457723	26.55928	115.2142 129.057
combined	189	129.8148	2.224323	30.57938	125.427 134.2027

```
    ratio = sd(0) / sd(1)                              f =    1.4267
Ho: ratio = 1                          degrees of freedom =   129, 58

   Ha: ratio < 1              Ha: ratio != 1               Ha: ratio > 1
 Pr(F < f) = 0.9356        2*Pr(F > f) = 0.1289         Pr(F > f) = 0.0644
```

Stata also provides so-called "immediate" commands, for which merely the data useful to build the test statistic are provided. In the case of the Student's *t*-test, the command of interest is ttesti whose signature is reproduced below:

```
ttesti #obs1 #mean1 #sd1 #obs2 #mean2 #sd2 [, options2]
```

This command thus expects the count, the mean and the standard deviation for the first sample and the same information for the second sample. The options (options2) correspond to the previously discussed options in the case of unequal variances (unequal or welch), as well as to the risk level $1 - \alpha$ (level).

2.1.2. *Non-independent samples*

In the case of two paired samples, the same principle as that evoked for two series of measures represented in the form of two variables will be employed (this time, the two variables have the same number of observations), that is a formulation of the type:

```
. ttest x1 == x2
```

where x1 and x2 represent the two paired series of measurement. It will be obviously assumed that the observations are arranged in the same order for the two variables. The option paired is optional in this case.

2.1.3. *Non-parametric approach*

The Wilcoxon test for independent samples, based on the ranks of the observations, is obtained with the command ranksum:

```
. ranksum lwt, by(low)

Two-sample Wilcoxon rank-sum (Mann-Whitney) test

         low |     obs    rank sum    expected
-------------+-----------------------------------
           0 |     130     13217.5       12350
           1 |      59      4737.5        5605
-------------+-----------------------------------
    combined |     189       17955       17955

unadjusted variance     121441.67
adjustment for ties       -181.97
                        ----------
adjusted variance       121259.70

Ho: lwt(low==0) = lwt(low==1)
           z =    2.491
   Prob > |z| =   0.0127
```

In the case of two paired samples, the signed rank test can be obtained with the command signrank and a syntax almost identical to that of the *t*-test for paired samples:

```
. signrank x1 = x2
```

2.2. Comparaisons of two proportions

2.2.1. Independent samples

A Pearson χ^2 test can be obtained in several ways, usually from commands that enable building a contingency table. When using `tabulate`, for example, the option `chi` will be added:

```
. tabulate low smoke, chi expected

+--------------------+
| Key                |
|--------------------|
|     frequency      |
| expected frequency |
+--------------------+

  Weight  |
less than |           smoker
  2.5 kg  |        0          1 |     Total
----------+----------------------+----------
        0 |       86         44 |       130
          |     79.1       50.9 |     130.0
----------+----------------------+----------
        1 |       29         30 |        59
          |     35.9       23.1 |      59.0
----------+----------------------+----------
    Total |      115         74 |       189
          |    115.0       74.0 |     189.0

          Pearson chi2(1) =   4.9237   Pr = 0.026
```

The option `expected` makes it possible to display the theoretical counts (expected under the hypothesis of independence between the two variables) in each cell of the table. With an option `exact`, Stata also returns the result of the Fisher's test.

The same result can be achieved by including the immediate command `tabi`. As in the case of the Student's t-test, it is necessary to provide the essential information for the construction of the test statistic. In the present case, it concerns the counts for the four cells in the contingency table, that is:

```
. tabi 86 44\ 29 30, chi2 exact nofreq

          Pearson chi2(1) =   4.9237   Pr = 0.026
```

```
           Fisher's exact =              0.036
   1-sided Fisher's exact =              0.020
```

It should be noted that the input of the frequency counts is carried out rowwise, separating the rows of the table with the symbol \.

With regard to the proportion tests for a sample (usually associated with the null hypothesis $H_0 : p = 0.5$), the binomial test is accessible via the command bitest (the corresponding immediate command is bitesti). If the intention is to use a normal approximation, the corresponding test is obtained with prtest (prtesti). This command also works in the case of two samples.

In the following, a sample application considers the distribution of the mothers according to the smoker status:

```
. bitest smoke == 0.5, detail

     Variable |      N   Observed k   Expected k   Assumed p   Observed p
-------------+---------------------------------------------------------------
        smoke |    189          74         94.5     0.50000     0.39153

  Pr(k >= 74)               = 0.998917  (one-sided test)
  Pr(k <= 74)               = 0.001754  (one-sided test)
  Pr(k <= 74 or k >= 115)   = 0.003508  (two-sided test)

  Pr(k == 74)               = 0.000671  (observed)
  Pr(k == 114)              = 0.001029
  Pr(k == 115)              = 0.000671  (opposite extreme)

. prtest smoke == 0.5

One-sample test of proportion                 smoke: Number of obs =      189
-------------------------------------------------------------------------------
     Variable |     Mean   Std. Err.                    [95% Conf. Interval]
-------------+-----------------------------------------------------------------
        smoke | .3915344   .0355036                     .3219487    .4611201
-------------------------------------------------------------------------------
    p = proportion(smoke)                              z =   -2.9823
Ho: p = 0.5

    Ha: p < 0.5                Ha: p != 0.5                Ha: p > 0.5
 Pr(Z < z) = 0.0014        Pr(|Z| > |z|) = 0.0029        Pr(Z > z) = 0.9986
```

In the first case, the quantity of interest is the one entitled `Pr(k <= 74 or k >= 115)`, and its equivalent is found in the proportion test based on the normal distribution under `Pr(|Z| > |z|)`. In the case of two samples, the following command allows to test the hypothesis that the distribution of the smoking mothers is the same regardless of the status of the baby's weight at birth (it is easy to compare the result of the two-tailed test with that of a χ^2 test):

```
. prtest smoke, by(low)

Two-sample test of proportions                   0: Number of obs =     130
                                                 1: Number of obs =      59
------------------------------------------------------------------------------
   Variable |      Mean   Std. Err.      z    P>|z|     [95% Conf. Interval]
------------+-----------------------------------------------------------------
          0 |  .3384615   .0415012                      .2571207    .4198024
          1 |  .5084746   .0650851                      .3809101     .636039
------------+-----------------------------------------------------------------
       diff |  -.170013   .0771908                     -.3213042   -.0187219
            |  under Ho:  .0766189   -2.22   0.026
------------------------------------------------------------------------------
       diff = prop(0) - prop(1)                              z =   -2.2189
   Ho: diff = 0

   Ha: diff < 0                 Ha: diff != 0                 Ha: diff > 0
 Pr(Z < z) = 0.0132       Pr(|Z| < |z|) = 0.0265         Pr(Z > z) = 0.9868
```

The commands `graph bar` (bar chart) or `graph dot` (dot plot) can display and summarize any count table. Figure 2.2 presents an example of bar chart, where we combine the approaches used p. 14 (creation of an auxiliary variable for the observations counts) and 26 (adding labels to the variables):

```
. replace freq = 1 // the variable already exists
(0 real changes made)
. graph bar (sum) freq, over(smoke, relabel(1 "Non smoking" 2 "Smoking")) ///
  asyvars over(low, relabel(1 "Normal" 2 "Low (< 2.5 kg)")) ///
  legend(title("Mother")) ytitle("Counts")
```

For such graphic representations, it is possible to install the package `catplot`, by typing:

```
. ssc install catplot
```

at the Stata command prompt. This assumes a functional internet connection. Providing a few options for customizing the chart (legend, labels for the variables modalities, etc.), the previous command appears as (Figure 2.3):

```
. catplot low smoke, recast(bar) ///
  var1opts(relabel(1 "Normal" 2 "Low (< 2.5 kg)"))
```

See the online help for more details: `help catplot`.

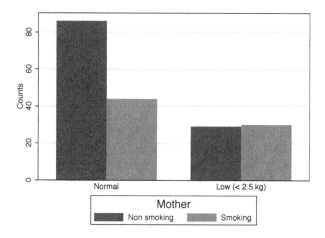

Figure 2.2. *Bar chart for two qualitative variables*

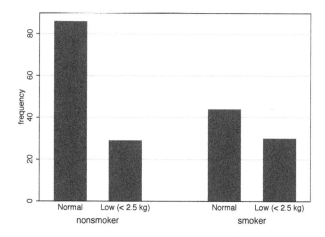

Figure 2.3. *Usage of catplot for the construction of a bar chart*

2.2.2. *Non-independent samples*

In the case of two binary variables observed for a same sample or two paired samples (for example, a case-control study), if the marginal frequencies of the contingency table resulting from the cross-tabulation of these two variables prove to be of interest, the McNemar test can be utilized with the command mcc. The syntax follows the notation commonly used in epidemiological studies (case/control, cohort studies) and Stata provides the exact degree of significance and that based on a $\chi^2(1)$ distribution. The immediate command is mcci.

2.3. Risk measures and OR

Most of the measures of risk or of association used in epidemiology can be found in a specific subset of Stata commands, known as epitab. They are also accessible through a specific menu, Statistics ▷ Epidemiology and related ▷ Tables for epidemiologists.

For example, the command tabodds is used in the case of case-control studies or cross-sectional studies. It allows the calculation of the OR and its asymptotic confidence interval, as well as testing the homogeneity of ORs between strata (Mantel–Haenszel test). Here follows an example of its usage where the status of the weight at birth (low) is considered as a response variable and the variable smoke as a risk or exposure factor:

```
. tabodds low smoke, or
```

```
---------------------------------------------------------------------------
    smoke |  Odds Ratio        chi2      P>chi2      [95% Conf. Interval]
----------+----------------------------------------------------------------
       0 |   1.000000           .           .            .           .
       1 |   2.021944         4.90       0.0269       1.069897    3.821169
---------------------------------------------------------------------------

Test of homogeneity (equal odds): chi2(1)  =      4.90
                                  Pr>chi2  =    0.0269

Score test for trend of odds:     chi2(1)  =      4.90
                                  Pr>chi2  =    0.0269
```

It is possible to instruct Stata what level of the second variable serves as the reference category by means of the option base(), and also to specify different confidence intervals (cornfield or woolf).

In the previous example, the individual data are available, but it often happens that the data are accessible in aggregated format, that is to say, in the form of a contingency table that can also be redrafted in the form of a three-entry table: the first two columns indicate the cross-tabulations of each level of the two binary variables and the third column indicates the associated frequency counts. The syntax remains the same in this case: it indicates the name of the variables acting as a response and as explanatory factors. Stata is, however, instructed how it should weight the processing (cross-tabulating the two modalities of each variable) with counts passed as an option `fweight=`. Assuming that the data be presented as shown in Table 2.1, the syntax would then be:

```
. tabodds low smoke [fweight=N], or
```

low	smoke	N
0	0	86
1	0	29
0	1	44
1	1	30

Table 2.1. *Frequency table for the cross-tabulation of the variables* low *and* smoke

Note that with this type of configuration (table with counts in a specific column), the graphics commands to create bar and dot plots are slightly simplified: it is no longer necessary to define a count variable and the option (sum) to accumulate the counts by variable levels. Furthermore, the count column can be regarded as a primary variable by specifying an option (asis).

Similarly, weighting by the frequency counts with the option [fweight=N] (that can be shortened to [fw=N]), where N refers to the variable containing the counts can be employed with the command `tabulate`. Such a table can be stored in a simple text file including the name of the three variables on the first line of the file (insheet will then be used to import it), or it is possible to directly input data with the command `input`. In this case, the operations can be performed between two commands preserve/restore to avoid losing the data of the current session. The user should bear in mind that the variables will still have to be renamed so as not to create conflicts with those present in the workspace. In the illustration here below, after the command `preserve` has been typed, the aggregated data table was created by appending the name of the variables low and smoke with b:

```
. input lowb smokeb N

    lowb  smokeb  N
 1. 0 0 86
 2. 1 0 29
 3. 0 1 44
```

```
  4. 1 1 30
  5. end

. list lowb-N in 1/4
      +--------------------+
      | lowb    smokeb   N |
      |--------------------|
  1. |   0        0    86 |
  2. |   1        0    29 |
  3. |   0        1    44 |
  4. |   1        1    30 |
      +--------------------+
```

Regarding the previous example, it thus should be written as:

```
. tabulate lowb smokeb [fw=N]

            |       smokeb
     lowb |       0         1 |     Total
-----------+----------------------+----------
        0 |      86        44 |       130
        1 |      29        30 |        59
-----------+----------------------+----------
    Total |     115        74 |       189
```

This example will serve as the background for illustrating a few customization options in dot plots, notably the specification of an axis with units chosen by the user and different symbols to highlight the levels of the classification variable (Figure 2.4):

```
. label define status 0 "Normal weight" 1 "Low weight"
. label define smoker 0 "Non smoker" 1 "Smoker"
. label values lowb status
. label values smokeb smoker
. graph dot (asis) N, over(smokeb) asyvars over(lowb) ///
    yscale(range(0 100)) ylabel(0(20)100) ///
    marker(1, msymbol(oh)) marker(2, msymbol(X))
```

It should be reminded that restore has to be typed to return to the initial data set.

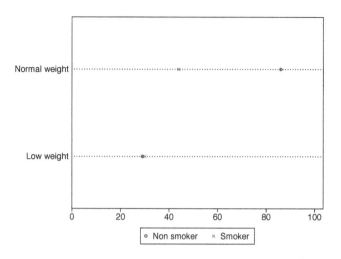

Figure 2.4. *Dot plot for the babies' weight distribution*

2.4. Analysis of variance

2.4.1. *One-way ANOVA*

The command `summarize` (in combination with `tabulate` or `by:`), or `tabstat`, allows for naturally summarizing the distribution of the response variable based on the levels of the explanatory variable. However, it will be seen that a numerical summary can be directly coupled to the command generating the ANOVA table. Concerning the graphical methods, it is always possible to utilize `graph bar` or `graph dot` to visually present the distribution of the group means; it will suffice to replace the summary statistics `r(sum)` by `r(mean)`. With respect to the distribution of the individual data, we can build counts (or proportions) histograms by means of the command `histogram`, indicating the classification factor in the option `by()`.

Consider the data related to the weight of the babies at birth (`bwt`) and the mothers' ethnic origin (`race`). The distribution of the weight, in terms of counts (option `freq`), can be obtained in the following manner (Figure 2.5):

```
. histogram bwt, by(race, col(3)) freq
```

The principal command to perform a one-way (fixed-effect) ANOVA is `oneway`. For more complex models, `anova` or `regress` will have to be used. Its usage is relatively simple: a list of variables is provided, in this case the response variable and then the explanatory variable. The option `tabulate` automatically adds a table summarizing group means and standard deviations to the ANOVA table:

```
. oneway bwt race, tabulate
```

```
              |        Summary of bwt
       race   |    Mean     Std. Dev.       Freq.
--------------+------------------------------------
      White   | 3102.7188    727.88615          96
      Black   | 2719.6923    638.68388          26
      Other   | 2805.2836    722.19436          67
--------------+------------------------------------
      Total   | 2944.5873    729.2143          189
```

```
                       Analysis of Variance
    Source              SS         df      MS            F      Prob > F
-----------------------------------------------------------------------
Between groups      5015725.25      2   2507862.63      4.91    0.0083
Within groups       94953930.6    186   510505.003
-----------------------------------------------------------------------
    Total           99969655.8    188   531753.488
```

```
Bartlett's test for equal variances:  chi2(2) =    0.6595  Prob>chi2 = 0.719
```

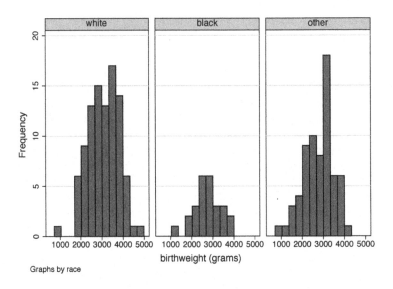

Graphs by race

Figure 2.5. *Distribution of the weight of newborn babies based on the mothers' ethnicity*

It should be noted that Stata also indicates the result of Bartlett's test for the equality of variances. If we want to apply the Levene test, the command `robvar` will have to be utilized, which returns the result under the test statistics named W0:

```
. robvar bwt, by(race)

             |         Summary of bwt
       race  |       Mean    Std. Dev.        Freq.
------------+-------------------------------------
      White  |   3102.7188    727.88615           96
      Black  |   2719.6923    638.68388           26
      Other  |   2805.2836    722.19436           67
------------+-------------------------------------
      Total  |   2944.5873     729.2143          189


W0  =  0.44717123   df(2, 186)      Pr > F = 0.64012002


W50 =  0.46842949   df(2, 186)      Pr > F = 0.62672105


W10 =  0.45725627   df(2, 186)      Pr > F = 0.63372775
```

The syntax is slightly different from that employed with the command `oneway`: this is a fully fledged command when testing for equality of variance (such as `sdtest`, see section 2.1.1), it is not directly connected to the ANOVA model built from `oneway`.

2.4.2. *Pairwise comparisons of means*

With regard to the comparisons of all pairs of means (three in the previous example), the simplest way consists of adding one of the correction options for multiple testing (`bonferroni`, `scheffe` or `sidak`) when making use of `oneway`. For the sake of clarity, the ANOVA table has not been displayed in the following expression:

```
. oneway bwt race, bonferroni noanova

                    Comparison of bwt by race
                          (Bonferroni)
Row Mean-|
Col Mean |      White        Black
---------+----------------------
   Black |   -383.026
         |      0.049
         |
```

```
Other |   -297.435    85.5913
      |     0.029     1.000
```

We will find the result for the comparison White *versus* Black by performing a simple *t*-test whose degree of significance is adjusted for all of the comparisons. The command `ttest` returns the test statistic ($r(t)$) and the *p*-value ($r(p)$), as it can be verified by typing `return list` after the first command:

```
. quietly: ttest bwt if race != 3, by(race)
. display r(p)*3
.04853058
```

2.4.3. *Linear trend test*

To perform a linear trend test, the linear regression approach is equivalent to replacing the command `oneway` by `regress`. For example, consider the variable `ftv`, which represents the number of visits to the gynecologist during the first trimester of pregnancy. This variable takes values between 0 and 6, values greater than 2 being rarely observed. This variable can be recoded in a three-class variable by means of the command `recode` whose syntax is quite simple: the new levels are indicated next to the old levels (the association is carried out with the symbol =) and the symbol / provides a way, as in the case of the operator `in`, to indicate a range of values (starting value/ending value):

```
. recode ftv (0=0) (1=1) (2/6=2), gen(ftv2)
(12 differences between ftv and ftv2)
```

The option `gen()` makes it possible to generate a new variable. It will be possible to verify that recoding has been correctly carried out with a simple cross-tabulation involving the two variables:

```
. tabulate ftv2 ftv
```

Without any other indication, the variable `race` will be regarded as a numeric variable, and considering that the distances between levels are equal (this is the case here because the levels are coded $\{1, 2, 3\}$), the test associated with the slope of the regression line provides the requested result:

```
. regress bwt ftv2
```

```
      Source |       SS       df       MS              Number of obs =     189
-------------+------------------------------           F(  1,   187) =    1.02
       Model |  542577.691      1  542577.691          Prob > F      =  0.3137
```

```
   Residual |  99427078.1    187   531695.605          R-squared      =  0.0054
------------+----------------------------------        Adj R-squared  =  0.0001
      Total |  99969655.8    188   531753.488          Root MSE       =   729.17


------------------------------------------------------------------------------
        bwt |     Coef.    Std. Err.      t    P>|t|     [95% Conf. Interval]
------------+-----------------------------------------------------------------
       ftv2 |   66.09496   65.42879     1.01   0.314    -62.97845     195.1684
      _cons |   2898.775   69.78422    41.54   0.000     2761.11      3036.441
------------------------------------------------------------------------------
```

The approach using contrasts to assess the linear trend test still relies on the use of the command `regress`, but this time Stata is required to treat the variable `ftv2` as a categorical variable by prefixing its name with the operator `i.` (see the online help, `help fvvarlist`). This has the effect of converting the variable with $k = 3$ levels into $k - 1 = 2$ dummy-coded variables, encoding for the levels j $(j = 2, \ldots, k)$ of the classification variable. Since the results of the regression on their own are not really of interest to us, their displaying is omitted by prefixing the regression command with the instruction `quietly:` and Stata is required to display the (orthogonal) polynomial contrasts associated with the explanatory variable. This last operation is achieved by prefixing the name of the grouping variable with the operator `p.`:

```
. quietly: regress bwt i.ftv2
. contrast p.ftv2, noeffects

Contrasts of marginal linear predictions

Margins      : asbalanced


--------------------------------------------------
             |     df          F        P>F
-------------+------------------------------------
        ftv2 |
    (linear) |      1        0.41     0.5216
 (quadratic) |      1        2.55     0.1119
       Joint |      2        1.79     0.1698
             |
    Residual |    186
--------------------------------------------------
```

The contrast of interest here is mentioned under the name (`linear`).

We will be able to verify the coefficient of determination, which is returned by the previous command, not as a result but as a postestimation (see `ereturn list`):

```
. display e(r2)
.01888498
```

which corresponds to the proportion of variance explained by the ANOVA model and that can be obtained as follows (rather than calculating it from the sums of squares displayed by oneway):

```
. anova bwt ftv2
```

```
                        Number of obs =      189    R-squared     =  0.0189
                        Root MSE      = 726.169    Adj R-squared =  0.0083

           Source |  Partial SS    df        MS             F     Prob > F
        ----------+----------------------------------------------------------
            Model |  1887925.36     2    943962.682         1.79     0.1698
                  |
             ftv2 |  1887925.36     2    943962.682         1.79     0.1698
                  |
         Residual |  98081730.4   186    527321.131
        ----------+----------------------------------------------------------
            Total |  99969655.8   188    531753.488
```

2.4.4. Computing specific contrasts

More generally, it is also possible to estimate, or to even test, any contrast with the command regress. To do this, the command lincom will be utilized. Here is an example with the initial model (babies' weight at birth and mothers' ethnicity):

```
. regress bwt i.race
```

```
       Source |       SS       df       MS                Number of obs =     189
    ----------+-----------------------------------         F(  2,   186) =    4.91
        Model |  5015725.25      2   2507862.63            Prob > F      =  0.0083
     Residual |  94953930.6    186   510505.003            R-squared     =  0.0502
    ----------+-----------------------------------         Adj R-squared =  0.0400
        Total |  99969655.8    188   531753.488            Root MSE      =   714.5


    ----------------------------------------------------------------------------
          bwt |     Coef.   Std. Err.      t    P>|t|     [95% Conf. Interval]
    ----------+-----------------------------------------------------------------
         race |
            2 |  -383.0264   157.9638    -2.42   0.016    -694.6575    -71.3954
```

```
     3  |  -297.4352    113.742    -2.61   0.010    -521.8254   -73.04498
        |
 _cons  |   3102.719   72.92298    42.55   0.000     2958.856   3246.581
------------------------------------------------------------------------
```

It can be seen that Stata provides a regression coefficient per level of the variable race, except for the first one which serves as the reference category. The y-intercept therefore represents the average weight of babies whose mothers are of the White type, and each of the two regression coefficients represents the deviation between the groups Black and Other with respect to the group White. The average difference between the two groups Black and Other can be estimated as follows (Stata "numbers" the regression coefficients based on the numerical codes of the factor levels, starting at 1):

. lincom 3.race - 2.race

(1)   - 2.race + 3.race = 0

```
-------------------------------------------------------------------------
    bwt  |    Coef.   Std. Err.      t    P>|t|    [95% Conf. Interval]
---------+---------------------------------------------------------------
    (1)  |  85.59127   165.0887    0.52   0.605    -240.0958   411.2783
-------------------------------------------------------------------------
```

Similarly, we know that the command ci can be used to form a confidence interval for an average employing the normal distribution (section 1.3), for example:

. lincom _cons + 1.race

(1)   1b.race + _cons = 0

```
-------------------------------------------------------------------------
    bwt  |    Coef.   Std. Err.      t    P>|t|    [95% Conf. Interval]
---------+---------------------------------------------------------------
    (1)  |  3102.719   72.92298    42.55   0.000    2958.856   3246.581
-------------------------------------------------------------------------
```

Care should be taken not to forget the y-intercept in the previous expression.

2.4.5. *Non-parametric approach*

The non-parametric alternative to ANOVA discussed above, known as Kruskal–Wallis ANOVA, is obtained with the command kwallis. The syntax is slightly different from that of oneway, anova or regress, and the classification factor appears this time in an option by(). Here are the results of the ANOVA based on the ranks with the same data of the weight at birth (bwt) and mothers' ethnicity (race):

```
. kwallis bwt, by(race)

Kruskal-Wallis equality-of-populations rank test

    +-----------------------+
    | race | Obs | Rank Sum |
    |------+-----+----------|
    | White | 96 | 10189.00 |
    | Black | 26 |  2015.00 |
    | Other | 67 |  5751.00 |
    +-----------------------+

chi-squared =      8.519 with 2 d.f.
probability =      0.0141

chi-squared with ties =     8.520 with 2 d.f.
probability =      0.0141
```

It is always possible to complete this analysis with pairwise comparisons for each level of the explanatory variable, via the command ranksum discussed in section 2.1.3. To isolate two groups among all of the groups of statistical units defined by the variable race, it suffices for example to exclude the third by means of a filter like if race != 3 (to compare groups White and Black, for example). A more economical alternative consists of installing the external command kwallis2 (type findit kwallis2 and follow the instructions for the installation). The syntax is identical to that of kwallis, but this command automatically provides all of the Wilcoxon tests associated with the model.

2.4.6. Two-factor ANOVA

ANOVAs with several classification criteria are performed with the command anova, more complex in usage than oneway but allowing that interaction terms be defined or specific contrasts tested. An example of its usage with the variables ht (history of hypertension in the mother) and race (mother's ethnicity), still considering infants' weight (bwt) as a response variable, is provided here below. A model including an interaction term will be considered, this latter being symbolized by ## in Stata. With the notation race##ht, Stata is required to consider two factors and their interaction (with oneway and anova, it is not necessary to inform Stata that variables must be explicitly represented in the form of categorical variables).

The results of the ANOVA model are indicated as follows:

```
. anova bwt race##ht
```

```
                        Number of obs =      189    R-squared     =  0.0768
                        Root MSE      = 710.143    Adj R-squared =  0.0516

        Source |  Partial SS    df       MS             F     Prob > F
    -----------+----------------------------------------------------------
         Model |  7682087.67     5   1536417.53         3.05    0.0115
               |
          race |  2992590.25     2   1496295.12         2.97    0.0539
            ht |  1757257.26     1   1757257.26         3.48    0.0635
       race#ht |  889649.132     2   444824.566         0.88    0.4157
               |
      Residual |  92287568.1   183   504303.651
    -----------+----------------------------------------------------------
         Total |  99969655.8   188   531753.488
```

In order to calculate sums of squares in sequential manner (as does the software R), it is imperative to add the option sequential.

An equivalent formulation of the model above that includes showing explicitly the two main effects and the interaction effect would be:

```
. anova bwt race ht race#ht
```

Numerical summary statistics can be built based on the same indicators (mean, standard deviation, etc.) that in the case where a single classification factor is under study. On the other hand, the command tabstat operates only with a classification variable. Furthermore, the command table can be employed as follows:

```
. table race, by(ht) contents(mean bwt sd bwt count bwt)
```

```
----------------------------------------------------
ht and     |
race       |  mean(bwt)    sd(bwt)     N(bwt)
-----------+----------------------------------------
No         |
     White |   3110.89    726.7152         91
```

```
  Black |    2751.83    542.1708          23
  Other |    2852.41    704.9773          63
----------+----------------------------------
Yes       |
  White |       2954    819.4086           5
  Black |    2473.33    1327.633           3
  Other |       2063    649.5717           4
------------------------------------------
```

or:

```
. table race ht, contents(mean bwt sd bwt count bwt)
```

An interaction graph can be constructed from the external command `anovaplot` (to be downloaded and installed, `findit anovaplot`), or from the graphics commands based on the command `margins` introduced in the recent versions of Stata. Here follows a possible solution (Figure 2.6):

```
. quietly: margins race#ht
. marginsplot
  Variables that uniquely identify margins: race ht
```

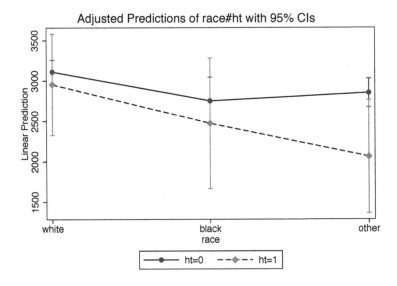

Figure 2.6. *Interaction diagram*

2.5. Key points

– The commands `ttest` and `ranksum` (or `signrank` paired samples) provide the comparison tests for two samples, whether independent or not, considering a numerical response variable.

– The commands `bitest` and `prtest` allow for the statistical testing of null hypotheses involving proportions calculated from one or two samples, by making use of the binomial or normal distribution.

– Most test commands for one or two samples can be applied from the summary statistics (average, standard deviation, proportion), without resorting to the full data: these are commands known as immediate.

– The command `tabulate` includes two options (`chi2` and `exact`) providing the χ^2 or Fisher statistics in the case of a contingency table, whereas the `epitab` commands such as `tabodds` allow calculating the OR, eventually taking into account a stratification factor.

– The commands relative to the ANOVA model are `oneway` (single classification factor case) and `anova` (ANOVA for multiple factors), and they include specific options (`bonferroni`, in the case of `oneway`) or have associated commands (`contrast`, in the case of `anova`) to perform pairwise comparisons, operate on specific contrasts or summarize marginal effects (useful for the construction of interaction graphs, for example).

2.6. Further reading

Mitchell's book [MIT 12] presents the commands `margins` and `marginsplot` in detail and discusses the utilization of contrasts in Stata with many illustrations. For an in-depth discussion of one- or two-way ANOVA, including the case of repeated measurements, the reader is invited to consult Dupont's work [DUP 09].

2.7. Applications

1) In this example, the quality of sleep in 10 patients has been measured before (control) and after treatment with one of the two following hypnotics: (1) D. hyoscyamine hydrobromide and (2) L. hyoscyamine hydrobromide. The evaluation criterion retained by the researchers was the average gain of sleep (in hours) compared to the basic duration of sleep (control) [STU 08]. The data are reported here below and are also present in the basic datasets of the R software (`data(sleep)`):

```
D. hyoscyamine hydrobromide:
0.7 -1.6 -0.2 -1.2 -0.1  3.4  3.7  0.8  0.0  2.0
L. hyoscyamine hydrobromide :
1.9  0.8  1.1  0.1 -0.1  4.4  5.5  1.6  4.6  3.4
```

The researchers have concluded that only the second molecule actually had a soporific effect.

– estimate the average sleeping time for each of the two molecules, as well as the difference between these two averages;

– display the distribution of the difference scores (LHH - DHH) in the form of a histogram, considering half-hour class intervals, and indicate the mean and the standard deviation of these difference scores;

– verify the accuracy of the findings by means of a Student's *t*-test.

The data from the study on sleep providing a basis for Student's article are not available in Stata. Nonetheless, we can manually input them with the command input, as was given on p. 20. In this case, we will therefore create two variables, DHH and LHH, corresponding to the sleep time recorded for the D. hyoscyamine hydrobromide and L. hyoscyamine hydrobromide molecules, respectively:

```
. input DHH LHH

       DHH  LHH
 1. 0.7 1.9
 2. -1.6 0.8
 3. -0.2 1.1
 4. -1.2 0.1
 5. -0.1 -0.1
 6. 3.4 4.4
 7. 3.7 5.5
 8. 0.8 1.6
 9. 0.0 4.6
10. 2.0 3.4
11. end
```

It is possible to verify that the input yields the expected results by displaying the first five observations:

```
. list in 1/5
```

Arithmetic means per treatment group are obtained using tabstat, without which another option returns the average of the variables listed in the command:

```
. tabstat DHH LHH, save
```

The option save mentioned above allows for the results returned by the command tabstat to be temporarily stored, which makes it possible to reuse them to calculate the average difference between the two molecules:

```
. return list
. matrix m = r(StatTotal)
. matrix list m
. display m[1,2] - m[1,1]
```

As has been seen with R, the manipulation of the auxiliary variable m, in which the two group averages have been stored, is achieved by calling its elements by position number (the average for the LHH group is in second position, it therefore can be used as m[1,2]).

The scores of differences will be calculated by simple subtraction, and the results will be stored in a new variable as shown below:

```
. gen sdif = LHH - DHH
. tabstat sdif, stats(mean sd)
```

Equivalently, we could use summarize sdif.

Finally, in order to display the distribution of the scores of differences in the form of a histogram imposing class intervals of 0.5 h, it will be necessary to add the options bin(8) (eight intervals in total) and start(0) (starting at 0):

```
. histogram sdif, percent bin(8) start(0)
```

To carry out a *t*-test for paired data, we always make use of the command ttest, this time with a slightly different syntax:

```
. ttest DHH == LHH
```

Finally, the average gains in sleeping time can be represented for each molecule by using a bar chart:

```
. graph hbar DHH LHH, bargap(20)
```

Here, the choice was made to represent the data in the form of horizontal bars (hbar instead of bar), knowing that by default Stata automatically calculates the conditional means. To display the median, the command graph hbar (median) DHH LHH would be employed.

2) In a clinical trial, the objective was to evaluate a therapy supposed to reduce the number of symptoms associated with benign breast diseases. A group of 229 women having this disease have been randomly divided into two groups. The first group received the routine care, while the patients in the second group followed a special therapy (variable B = treatment). After 1 year, the individuals have been evaluated and have been classified in one of two categories: improvement or no improvement (variable A = response) [SEL 98]. The results are summarized in Table 2.2, for a portion of the sample.

	therapy	no therapy	total
improvement	26	21	47
no improvement	38	44	82
total	64	65	129

Table 2.2. *Therapy and breast disease*

– perform a chi-square test;

– what are the theoretical counts expected under an independence hypothesis?

– compare the results obtained using a chi-squared test with those of Fisher's test;

– give a confidence interval for the difference in improvement proportion between the two groups of patients.

Stata includes commands called "immediate" that enable the calculation of the test statistics associated with data directly entered at the command prompt, that is without having to import a data file or to input a counts table. In this case, an answer can be provided to the three questions raised with the same command, tabi (not to be confused with the external command chitesti that carries out testings on fitting to a particular distribution):

```
. tabi 26 21\ 38 44, exact chi2 expected
```

The option expected makes it possible to provide the theoretical counts, in parallel to the observed counts.

NOTE.– Stata does not provide Yates's correction for continuity for this type of statistical testing among the basic commands.

With regard to the rate of improvement in both groups, it is possible to utilize prtesti in the following manner:

```
. prtesti 64 0.233 65 0.237
```

3) In studies on estrogen receptor genes, geneticists have directed their interest toward the relationship between the genotype and the age of diagnosis of breast cancer. The genotype was determined from the two alleles of a sequence restriction polymorphism (1.6 and 0.7 kb), that is three groups of subjects: homozygous patients for the allele 0.7 kb (0.7/0.7), homozygous patients for the allele 1.6 kb (1.6/1.6) and heterozygous patients (1.6/0.7). The data have been collected on 59 patients with breast cancer and are available in the file polymorphism.dta (Stata file) [DUP 09]. The average data are shown in Table 2.3.

	Genotype			Total
	1.6/1.6	1.6/0.7	0.7/0.7	
Number of patients	14	29	16	59
Age at diagnostic				
Average	64.64	64.38	50.38	60.64
Standard deviation	11.18	13.26	10.64	13.49
95% CI	(58.1-71.1)	(59.9-68.9)	(44.3-56.5)	

Table 2.3. *Estrogen receptor gene polymorphism*

– test the null hypothesis according to which the age during diagnosis does not vary according to the genotype by means of an ANOVA. Represent the distribution of the ages for each genotype in a graphical form;

– the confidence intervals presented in Table 4.1 have been estimated on the assumption of the homogeneity of variances, that is to say, using the estimate of the common variance; give the value of these confidence intervals without assuming any homoscedasticity;

– estimate the differences in average corresponding to all possible combinations of the three genotypes, with an estimate of the associated 95% confidence interval and a parametric test for assessing the degree of significance of the observed difference;

– graphically represent the averages of groups with 95% confidence intervals.

Loading data is not really problematic since the data are already available in Stata format. To provide a numerical summary of the variable `age` for each genotype, a combination of the commands by and summarize can be used. It will later be seen that the command tabstat allows that simplified results be returned, for example only the mean and standard deviation:

```
. use polymorphism.dta
. by genotype: summarize age
```

The confidence intervals reported above are not calculated from the ANOVA performed here below, but considering an approximation with the normal distribution. However, a more detailed description summary can be obtained by adding the option `detail`:

```
. by genotype: summarize age, detail
```

It is possible to display the distribution of the ages for each genotype by means of box plots:

```
. graph box age, over(genotype)
```

When the distribution of the ages is to be represented with histograms, the command `histogram` can be used as follows:

```
. histogram age, by(genotype, col(3))
```

The option `by(genotype, col (3))` allows displaying the distributions conditionally to the genotype and that the histograms be horizontally aligned (that is using three columns).

The command `oneway` makes it possible to carry out a fixed effect analysis of variance as follows:

```
. oneway age genotype
```

The ANOVA table produced by Stata is substantially identical to that generated by R, except that Stata also returns the totals for the associated sums of squares, mean squares and degrees of freedom. It should also be noted that Stata provides the result of a Bartlett's test concerning the hypothesis of homogeneity of the variances. The Levene's test for testing the homogeneity of variances can be obtained employing a somewhat different approach: the command `robvar` that provides a collection of robust tests for the equality of variances:

```
. robvar age, by(genotype)
```

The Levene's test is presented under the test statistic W0.

Finally, note that the numerical summary resulting from the previous step (conditional means) can be partially reproduced (without the confidence intervals) adding the option `tabulate`:

```
. oneway age genotype, tabulate
```

In order to compute confidence intervals based on the results of the ANOVA, that is considering an estimation of the residual error based on the common variance, we can proceed as in the case of R, by utilizing the mean square of the error (`Within groups`) and the reference quantile of the distribution t (0.975) that can be obtained as follows in Stata (here for the first genotype, 1.6/1.6):

```
. display invttail(14-3, 0.025)
```

More generally, we could generate the upper and lower bounds of the 95% confidence interval of three group means from the aggregated data, as shown in the following example (in this case, the estimate of the common variance is no longer taken into account):

```
. collapse (mean) agem=age (sd) ages=age (count) n=age, by(genotype)
. generate agelci = agem - invttail(n-1, 0.025)*(ages/sqrt(n))
. generate ageuci = agem + invttail(n-1, 0.025)*(ages/sqrt(n))
```

and a command such as twoway (bar agem genotype) (rcap ageuci agelci genotype) would make it possible to display the results graphically. An alternative and more convenient solution is to use the command serrbar that can display a series of averages associated with their standard errors. Since we would use confidence intervals rather than standard errors, it just suffices to add a small modification to the standard usage. By employing the previous calculations, the half-width of the CI can be calculated (knowing that it is symmetric around the average):

```
. gen ageb = agem-agelci
. serrbar agem ageb genotype, xlabel(1 "1.6/1.6" 2 "1.6/0.7" 3 "0.7/0.7")
```

It should be noted that the differences in averages can be obtained by forming specific contrasts based on a regression model that gives results equivalent to the ANOVA model subject to encoding the categorical variable genotype in the form of a matrix of indicators by means of the operator i.*. We do not present the results of the call to the command regress since what matters to us is simply to make use of the results it saves (see e()):

```
. regress age i.genotype
```

For the difference in averages between genotypes 0.7/0.7 and 1.6/0.7, for example, the following command should be employed:

```
. lincom 3.genotype - 2.genotype
```

The same procedure will be followed for the other two average differences, 0.7/0.7 - 1.6/1.6 and 1.6/0.7 - 1.6/1.6.

In fact, the same approach (by regression) would allow us to obtain the 95% confidence intervals for each group average:

```
. lincom _cons + 1.genotype
```

4) Any obstetrics service has interest on the weight of term and 1 month-old infants [PEA 05]. With regard to this sample of 550 babies, there is also information available concerning the parity (number of brothers and sisters), but it is known that there is no twinhood relationship among children having brothers and sisters. The purpose of the study is to determine whether the parity (four classes) influences the weight of 1-month-old newborn babies. The data are summarized in Table 2.4 and they are available in an SPSS file named weights.sav.

– verify the data exposed in the previous table;

– perform the one-way ANOVA. Draw conclusions about the global significance and indicate the part of variance explained by the model;

– display the weight distribution according to gender. Perform a test for homogeneity of variance (search in the online help for Levene's test);

– it was decided that the last two categories be combined (2 and $\geq$ 3). Carry out the analysis once more and compare with the results obtained from the first ANOVA;

– perform a linear trend test (with ANOVA) with the data recoded into three parity levels.

	Number of brothers and sisters				Total
	0	1	2	$\geq$ 3	
Sample					
Count	180	192	116	62	550
Proportion	32.7	34.9	21.1	11.3	100.0
Weight (kg)					
Mean	4.26	4.39	21.1	11.3	
Standard deviation	0.62	0.59	0.61	0.54	
(Min-Max)	(2.92-5.75)	(3.17-6.33)	(3.09-6.49)	(3.20-5.48)	

Table 2.4. *Newborns' weight*

There is no very convenient solution to directly import an SPSS file in Stata, except when working on Windows (see the command usespss). The file containing the data, weights.sav, has been exported from R in the Stata format with the following commands:

```
library(foreign)
weights <- read.spss("weights.sav", to.data.frame=TRUE)
write.dta(weights, file="weights.dta")
```

Inside Stata, it can then very simply be imported inserting the command use:

```
. use "weights.dta", clear
. list in 1/5
```

The counts and relative frequencies table for the variable PARITY is obtained with the command tabulate:

```
. tabulate PARITY
```

The means and standard deviations of the weight according to the number of siblings are thus obtained:

```
. tabstat WEIGHT, stats(mean sd) by(PARITY) format(%9.2f)
```

The option format (%9.2f) limits the display to two decimal numbers.

A single-factor ANOVA is achieved as in the previous exercises, by means of the command oneway and by providing the response variable and the categorical variable describing the groups to be compared:

```
. oneway WEIGHT PARITY
```

It is also possible to calculate the proportion of variance explained based on the sums of squares previously reported. It is also possible to employ the command anova which returns, in addition to an ANOVA table, the coefficient of determination associated with the model:

```
. anova WEIGHT PARITY
```

In order to display the data in the form of histograms for each group, the command histogram has to be used with the option by to define the classification factor. The option freq allows counts to be displayed rather than proportions (or densities):

```
. histogram WEIGHT, by(PARITY) freq
```

To display a scatter plot, as in R, we can enter an external command, such as stripplot (ssc install stripplot), for example:

```
. stripplot WEIGHT, over(PARITY) stack height(.4) center vertical width(.3)
```

or more simply:

```
. scatter WEIGHT PARITY, jitter(3) xlabel(1 "Singleton" 2 "One sibling" 3
        "2 siblings" 4 "3 more")
```

By default, the command oneway displays the result of a Bartlett's test to compare the variances of the groups between them. If a Levene's test is to be carried out, the command robvar has to be used which yields the result of the Levene test (W0):

```
. robvar WEIGHT, by(PARITY)
```

There are several ways to recode variables in Stata, but in this case the simplest manner to aggregate the last two classes (2 siblings and 3 or more) consists in generating a second variable, PARITY2, as follows:

```
. recode PARITY (1=1) (2=2) (3/4=3), gen(PARITY2)
```

To verify that the conversion has successfully proceeded, a simple contingency table will be displayed crossing the counts of the two variables:

```
. tabulate PARITY PARITY2
```

The results of the one-way ANOVA considering the newly created variable are reported below:

```
. oneway WEIGHT PARITY2
```

For the linear trend test, two ways are presented (contrasts method and linear regression). To generate a contrast testing the linear trend, we must explicitly instruct

Stata to perform a regression by considering the variable PARITY2 as a qualitative variable (involving dummy variables coding for the two last factor levels). In this case, the post-estimation command contrast will be employed. Since the result of the régression on the categorical variable does not really interest us, Stata will be informed not to display the results with the instruction quietly:

```
. quietly: regress WEIGHT i.PARITY2
. contrast p.PARITY2, noeffects
```

The contrast of interest appears on the row entitled (linear).

With regard to the simple linear regression approach, the command is simpler:

```
. regress WEIGHT PARITY2
```

The trend test corresponds to the test associated with the slope of the regression line, here the coefficient PARITY2.

5) The data set ToothGrowth available in R contains data from a study on the length of odontoblasts (variable len) in 10 Guinea pigs after administration of different doses of C vitamin (0.5, 1 or 2 mg, variable dose) in the form of ascorbic acid or orange juice (variable supp) [BLI 52].

Data are available in the file ToothGrowth.csv

– verify the distribution of counts according to the different treatment conditions (cross-tabulating the modalities of the two factors, supp and dose) of this experimental design;

– calculate the mean and the standard deviation of each treatment;

– build a table of the analysis of variance for the full model with interaction between the two factors;

- draw an interaction chart that represents the average values of the response variable according to the levels of the two explanatory variables;

– verify that the homogeneity of variances assumption is acceptable for these data.

Loading data is carried out with the command import delimited since this is a regular CSV file. Assuming that the file is located in the working directory, we will thus type:

```
. import delimited "ToothGrowth.csv"
```

It should be noted that the variable supp is recognized as a character string. The distribution of statistical units according to the different processes is obtained with tabulate, but it might as well be possible to combine the calculation of averages and standard deviations with the enumeration of counts by means of table as illustrated below:

```
. table supp dose, contents(mean len sd len N len) format(%5.1f)
```

The instruction format(%5.1f) allows us to limit the output to one decimal place. Two-way ANOVA is performed with anova, but it is necessary to prior convert the variable supp into a categorical variable. To do this, encode is employed; this enables the conversion of strings into categorical variable levels (caution, it is necessary to create a new variable):

```
. encode supp, gen(suppc)
```

The variable dose is manipulated as a numeric variable but here we would like to convert it also into a categorical variable. This can be achieved in two stages: firstly, the variable is converted into a string of characters, and then it is transformed into a categorical variable as it was done for supp:

```
. tostring dose, generate(dose_)
. encode dose_, gen(dosec)
. drop dose_
```

The ANOVA model is then written:

```
. anova len suppc##dosec
```

The operator ## represents the interaction between the two factors. It should be noted that it is still possible to manipulate the variable dose as a numeric variable (1 degree of freedom in the ANOVA table), but in this case it is necessary to prefix it with the operator c. (anova len suppc##c.dose).

To construct an interaction graph, we can enter the command scatter, but the command marginsplot is also flexible in usage. It just suffices to initially calculate the conditional averages by means of margins as shown below:

```
. margins dosec#suppc
. marginsplot
```

Linear Regression

This chapter is dedicated to the measures of association between two numeric variables (linear correlation and non-parametric correlation) and to the linear regression model: estimation of the regression coefficients, fitting and prediction from new data with confidence intervals.

3.1. Measures of association between two numeric variables

3.1.1. *Bivariate descriptive statistics*

The command summarize provides information about the central tendency, the dispersion and the range of the values observed for a list of variables. It is therefore quite possible to use it to summarize the distribution of two numeric variables. In the case of the weight of babies (bwt) and mothers (lwt), the following results are available (the mothers' weights are expressed in pounds, and they are initially converted in kilograms):

```
. replace lwt = lwt/2.2
. summarize bwt lwt

lwt was int now float
(189 real changes made)

    Variable |       Obs        Mean    Std. Dev.       Min        Max
-------------+--------------------------------------------------------
         bwt |       189    2944.587    729.2143        709       4990
         lwt |       189    59.00673    13.89972   36.36364   113.6364
```

Note that the option detail allows more detailed information to be provided (quartiles, skewness and kurtosis, etc.).

Before calculating any linear or monotonic measure of association, it is advised that one should visually inspect a scatter plot that describes the covariation between the two sets of measurements. To do this, we use the command `scatter` specifying the two variables of interest: the first will be reported on the y-axis and the second on the x-axis:

```
. scatter bwt lwt
```

Superimposing a lowess curve on the previous graph gives an idea of the linearity of the relationship between the two variables and possible local deviations. To do this, it suffices to combine two `twoway` commands, delimiting them with parentheses. In the following case, the first command (`scatter bwt lwt`) displays the scatterplot, and the second (`lowess bwt lwt`) displays a lowess curve (Figure 3.1). Note that, `lowess` could be replaced by `lfit` to draw the regression line:

```
. twoway (scatter bwt lwt) (lowess bwt lwt), legend(off) ytitle("bwt")
```

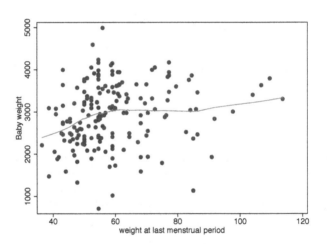

Figure 3.1. *Babies' (g) and mothers' weight (kg)*

Note that with the recent versions of Stata, the previous expression can be formulated:

```
. scatter bwt lwt || lowess bwt lwt, legend(off) ytitle("bwt")
```

In this case, the operator `||` is employed to separate the multiple graphics instructions.

In the case where more than two variables are involved, the command graph matrix, followed by the list of variables of interest, makes it possible to display all the scatterplots cross-tabulating the variables two by two.

3.1.2. *Pearson's correlation*

The command providing the Pearson's correlation coefficient between two numeric variables is correlate:

```
. correlate lwt bwt

(obs=189)

             |      lwt       bwt
-------------+------------------
         lwt |   1.0000
         bwt |   0.1857    1.0000
```

Note that an option means could be added to the previous command in order to simultaneously display the results produced by summarize. Normally reserved for the case of correlations between more than two variables, it is also possible to employ the command pwcorr. Nonetheless, the latter has the advantage of providing the results of the test of the null hypothesis for the parameter of interest:

```
. pwcorr lwt bwt, obs sig

             |      lwt       bwt
-------------+------------------
         lwt |   1.0000
             |
             |      189
             |
         bwt |   0.1857    1.0000
             |   0.0105
             |      189       189
             |
```

The command corrci provides an estimate of the confidence interval for the Pearson coefficient (by default, by making use of the inverse Fisher transformation). The option level() allows the desired confidence level to be modified:

```
. corrci lwt bwt
```

```
(obs=189)

              correlation and 95% limits
lwt bwt       0.186    0.044    0.320
```

3.1.3. *Non-parametric correlation*

When it is preferable to work with a measure of correlation based on the ranks of the observations, the Spearman coefficient of correlation will be employed with the command spearman:

```
. spearman lwt bwt

Number of obs =       189
Spearman's rho =        0.2489

Test of Ho: lwt and bwt are independent
     Prob > |t| =        0.0006
```

The command spearman also works with more than two variables and returns, such as pwcorr, a matrix of correlation coefficients with eventually the degree of significance (adding the option stats(rho p)).

3.2. Linear regression

3.2.1. *Estimation of the model parameters*

In Chapter 2, we have seen that the command regress (p. 40) is utilized when modeling the relationship between two variables, one being considered as a response variable. In this case, to model the relationship between the babies' weight (response variable) and the mothers' weight (explanatory variable), the following syntax would be used:

```
. regress bwt lwt

      Source |       SS      df       MS              Number of obs =     189
-------------+------------------------------          F(  1,   187) =    6.68
       Model |   3448639.3      1   3448639.3          Prob > F      =  0.0105
    Residual |  96521016.5    187  516155.168          R-squared     =  0.0345
-------------+------------------------------          Adj R-squared =  0.0293
       Total |  99969655.8    188  531753.488          Root MSE      =  718.44
```

```
--------------------------------------------------------------------------------
        bwt |    Coef.    Std. Err.      t     P>|t|     [95% Conf. Interval]
------------+-------------------------------------------------------------------
        lwt |  9.744038   3.769686    2.58    0.011     2.307461    17.18061
      _cons |  2369.623   228.4932   10.37    0.000     1918.868    2820.379
--------------------------------------------------------------------------------
```

For any regression model in Stata, the variable response comes first, followed by the explanatory variable(s). This command provides the analysis of variance (ANOVA) table for the regression as well as the regression coefficients table (with confidence intervals), whose displaying may be omitted by including the option notable. It is always possible to redisplay the results of the regression model by simply typing the name of the command, regress (this is valid for the other regression models in Stata).

It is also possible to store or display the regression coefficient for the variable lwt only (slope of the regression line) by operating on the values returned by Stata. For example, the following statement displays the requested result:

```
. display _b[lwt]
9.7440378
```

This is a so-called *post-estimation* result, as discussed in Chapter 2. The command ereturn list provides the list of the postestimation values stored by Stata. In this case, the regression coefficients (y-intercept and slope) are stored in an object called e(b). If the previous instruction regress is slightly modified, it can be verified that these values are individually accessible:

```
. regress bwt lwt, noheader coeflegend
```

```
--------------------------------------------------------------------------------
        bwt |    Coef.   Legend
------------+-------------------------------------------------------------------
        lwt |  9.744038   _b[lwt]
      _cons |  2369.623   _b[_cons]
--------------------------------------------------------------------------------
```

The value of the coefficient of determination can also be displayed by making use of e(r2). In the following illustration, we insert the command display with a combination of text and numeric value (rounded to two decimal places):

```
. display "Coefficient of determination = " %3.2f e(r2)*100 " %"
Coefficient of determination = 3.45 %
```

When using display, the numerical results displayed on-screen can be formatted by prefixing the name of the variable whose content is to be displayed by a formatting instruction %x.yf (x values, including y decimal places).

The regression line can be represented graphically by means of the command twoway lfit bwt lwt, but as was mentioned in the previous section, this command can be combined with a simple scatter plot, for instance:

```
. twoway (scatter bwt lwt) (lfit bwt lwt)
```

3.2.2. *Pointwise and interval-based prediction*

In Stata, the general principle consists of estimating the parameters of a regression model to then proceed with postestimation commands. This is valid for the calculation of the predicted values or residuals of the regression model. When we want to calculate the fitted values for the model (predicted values of bwt for the observed values of lwt), we will make use of the command predict after an estimation command such as regress. The predicted values will always correspond to the latest regression model:

```
. predict double p, xb
```

The option xb (which is the default) provides the fitted values of the previous model. It is important to remember that a variable name be specified, here p, to store the predictions! The option double makes it possible to restrict the size of the memory storage of the predicted values.

The previous command does not provide any confidence intervals. However, it is not difficult to obtain the standard error for the fitted values and calculate from predicted values the associated confidence intervals. Take the case of the fitted values:

```
. predict sep, stdp
. generate lci = p - 1.96*sep
. generate uci = p + 1.96*sep
```

The variables sep, lci and uci correspond to the standard error and the upper and lower bounds of the 95% CI, respectively. These values could be used to manually display the regression line and its 95% confidence interval (the command line allows lines to be drawn in Stata), but it is easier and faster to use the command lfitci as shown hereafter (Figure 3.2):

```
. twoway (lfitci bwt lwt) (scatter bwt lwt)
```

The option `stdp` will be replaced by `stdf` to obtain the standard error for the prediction of new observations.

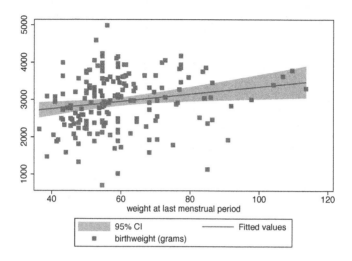

Figure 3.2. *Regression line (and 95% intervals) for the relation between the babies' weight (g) and the mothers' weight (kg)*

3.2.3. Model diagnostic

The command `estat` provides a certain amount of information concerning the goodness of fit of the model and helps to diagnose potential collinearity problems (`estat vif`) in the case where the model includes several explanatory variables:

```
. estat ic

-----------------------------------------------------------------------
  Model |   Obs    ll(null)   ll(model)     df         AIC         BIC
--------+--------------------------------------------------------------
      . |   189    -1513.56    -1510.242      2    3024.485    3030.968
-----------------------------------------------------------------------
           Note:  N=Obs used in calculating BIC; see [R] BIC note
```

It is also possible to use the external command `fitstat` (to be installed from the internet, `findit fitstat`) for a more detailed summary of the goodness of fit of the model:

```
. fitstat
```

```
Measures of Fit for regress of bwt

Log-Lik Intercept Only:     -1513.560   Log-Lik Full Model:       -1510.242
D(187):                      3020.485   LR(1):                         6.635
                                        Prob > LR:                     0.010
R2:                              0.034   Adjusted R2:                   0.029
AIC:                            16.003   AIC*n:                      3024.485
BIC:                          2040.278   BIC':                         -1.393
BIC used by Stata:            3030.968   AIC used by Stata:          3024.485
```

To obtain the residuals of the regression model (difference between the observed values and those predicted according to the model), we always make use of the command predict, specifying this time an option among: residuals (raw residuals), rstandard (standardized residuals) or rstudent (studentized residuals). In the series of instructions that follows, the three types of residuals are calculated, and their numerical summary is displayed using summarize with values rounded to five decimal places:

```
. predict double r, rstandard
. predict double rr, residuals
. predict double rrr, rstudent
. format r-rrr %9.5f
. summarize r-rrr, format
```

```
    Variable |      Obs       Mean    Std. Dev.        Min        Max
-------------+-----------------------------------------------------------
         r |      189   -0.00044      1.00210   -3.06017    2.89709
        rr |      189   -0.00000    716.52611  -2.19e+03   2.08e+03
       rrr |      189   -0.00107      1.00806   -3.13139    2.95644
```

In order to visualize the distribution of the residuals, a counts histogram can be employed by means of the command histogram, or a density curve representation. Figure 3.3 provides an illustration:

```
. kdensity r
```

It is also interesting to look at the distribution of the residuals with respect to the predicted values to verify the consistency of the variance and the absence of specific variation patterns in the residuals. To do this, it suffices to combine two twoway commands, on the same principle as in section 3.1.1 (Figure 3.4):

```
. twoway (scatter r p) (lowess r p), yline(0, lcolor(black) lpattern(dash))
         legend(off)
```

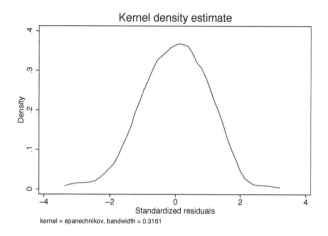

Figure 3.3. *Distribution of the residuals for the regression model of the babies' weight according to the mothers' weight*

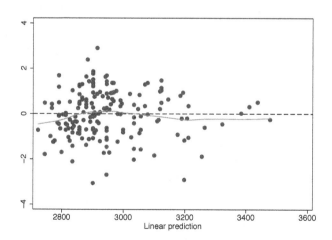

Figure 3.4. *Residuals and predicted values by the regression model*

In fact, the same graph can be obtained directly employing the command rvfplot that provides a representation of the residuals according to the predicted values (Figure 3.5):

```
. rvfplot, mlabel(smoke)
```

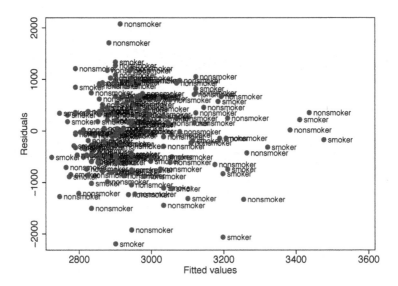

Figure 3.5. *Residuals and predicted values by the regression model including individuals annotation*

3.3. Multiple linear regression

The extension to the multiple regression model does not really raise any problems from the point of view of the instructions: the command `regress` receives the name of the response variable followed by all the explanatory variables. As it is often useful to transform certain variables, or to center them on their average, we will use this section to indicate how to center an explanatory variable. Knowing that when entering `summarize` this command generates a certain amount of information that can be used after (see `return list`), it is not difficult to center the variable relative to the mothers' weights on its average by proceeding as indicated in the following way:

```
. quietly: summarize lwt
. generate lwts = lwt - r(mean)
```

Note that if one intended to standardize the variable `lwt` (that is not only to subtract the average to each observation but also to normalize by the standard deviation), the expression above should be replaced by:

```
. generate lwts = (lwt - r(mean)) / r(sd)
```

By making use of the commands egen, the same result could be accomplished with a command such as:

```
. egen lwts = std(lwt)
```

Finally, the regression model, including the mothers' weight centered on their average and the frequency of visits to the gynecologist in the first trimester of pregnancy, would be written as:

```
. regress bwt lwts ftv i.race, noheader
```

```
--------------------------------------------------------------------------------
        bwt |      Coef.   Std. Err.       t    P>|t|     [95% Conf. Interval]
------------+-------------------------------------------------------------------
       lwts |   10.14028    3.891895     2.61   0.010      2.461807    17.81876
        ftv |   11.82137    49.17071     0.24   0.810     -85.18953    108.8323
            |
       race |
          2 |  -450.1195    158.1307    -2.85   0.005     -762.1019   -138.1371
          3 |  -239.2497    114.4963    -2.09   0.038     -465.1442   -13.35526
            |
      _cons |    3081.94    83.91919    36.73   0.000      2916.372    3247.507
--------------------------------------------------------------------------------
```

Note that the operator i. has been included to inform Stata to manipulate the variable race as a categorical variable, as indicated in Chapter 2 (section 2.4).

3.4. Key points

– The correlate and spearman commands enable the calculation of the Pearson and the Spearman correlation coefficients. A scatter diagram can be constructed with twoway scatter in order to visualize the shape of the cloud of points.

– The regress command is employed in the context of the linear regression and follows the same notation as in the case of the ANOVA (response variable followed by the explanatory variables).

– A number of postestimation commands allow for information to be obtained about the goodness of fit of the model (fitstat). These commands also make it possible to calculate the fitted values or to predict the expected values for new observations (predict).

3.5. Further reading

As in Chapter 2, Mitchell's book [MIT 12] provides numerous graphic illustrations (command `marginsplot`) applied to the case of the linear model. The Stata applications proposed by Vittinghoff *et al.* [VIT 05] contribute to deepening the key concepts in the construction and the interpretation of a multiple regression model.

3.6. Applications

1) In [EVE 01], the authors focused on measuring malnutrition among 25 patients aged from 7 to 23 years and suffering from cystic fibrosis. Information of the anthropometric characteristics and pulmonary function of these patients was included in the calculations (height, weight, etc.). The data are available in the file `cystic.dat`.

– calculate the linear correlation coefficient between the variables PEmax and Weight, as well as its 95% confidence interval;

– display the totality of the numerical data in the form of scatterplots, that is 45 graphs arranged in the form of a "dispersion matrix";

– calculate all of the Pearson and Spearman correlations between the numerical variables;

– calculate the correlation between PEmax and Weight, controlling the age (Age) (partial correlation). Graphically represent the covariation between PEmax and Weight highlighting the two most extreme terciles for the variable Age.

Tabulated data are available in a text file that can be imported by entering the command `insheet`. However, it is also possible to import the data with `infile`, which has the advantage of operating even when the data do not originate from an Excel-like table:

```
. infile int (Sub Age Sex) Height Weight BMP FEV RV FRC ///
  TLC PEmax using "cystic.dat" in 2/26, clear
```

In the above instruction, it is explicitly required that Stata encode the first three variables as integers and not as real numbers, which does not currently improve much but makes it possible to save memory space for large data sets. As was done with R, the variable Sex will be recoded in the first instance into a categorical variable to avoid any confusion:

```
. label define labsex 0 "M" 1 "F"
. label values Sex labsex
. tabulate Sex
```

The estimation of the Pearson correlation coefficient is carried out with the command `correlate`, but this does not provide any options for the confidence interval. External commands such as `ci2` can be employed that work on the same principle as the internal command `ci` (for averages and proportions) in the parametric or non-parametric case (Spearman) or either `corrci`, only in the parametric case:

```
. correlate PEmax Weight
. corrci PEmax Weight
```

To test this correlation coefficient against the hypothesis $H_0 : \rho = 0$, the command `pwcorr` can be entered with the option `sig`. In the case of several numerical variables, this command also allows the correlation matrix to be calculated as will be mentioned hereafter:

```
. pwcorr PEmax Weight, sig
```

To display the totality of the scatterplots, the `graph matrix` command is employed specifying the list of variables that we would like to present in the chart. Here, the variable `sex` will be omitted:

```
. graph matrix Age Height-PEmax
```

The correlations for each pair of variables are obtained with the `pwcorr` command, and the same notation will be used to indicate the variables that interest us:

```
. pwcorr Age Height-PEmax
```

Concerning Spearman correlations, `pwcorr` will be replaced by `spearman`:

```
. spearman Age Height-PEmax, stats(rho)
```

The partial correlation between the `PEmax` and `Weight` variables taking `Age` into account is obtained by means of the command `pcorr`; the variable of interest must be located in the first position in the list of variables and by default Stata displays all partial correlation coefficients for the other variables:

```
. pcorr PEmax Weight Age
```

In order to compute terciles of the `Age` variable, an auxiliary variable can be created with `egen` and the function `cut`. But we can directly utilize the `xtile` command whose usage is easier:

```
. xtile Age3 = Age, nq(3)
```

Finally, for scatterplots, here is a first solution:

```
. scatter PEmax Weight if Age3 != 2, mlab(Age3)
```

We could also (second solution) perform two calls to the scatter command, one following the other, each time restricting the sample to the sole classes of Age3 that are of interest (1 and 3, in this case):

```
. scatter PEmax Weight if Age3 == 1, msymbol(circle) || scatter PEmax ///
  Weight if Age3 == 3, msymbol(square) ///
  legend(label(1 "1st tercile") label(2 "3rd tercile"))
```

2) In the Framingham study, we can access data about systolic blood pressure (sbp) and about the body mass index (bmi) of 2,047 men and 2,643 women [DUP 09]. The main interest lies in the relationship between these two variables (after logarithmic transformation) in men and women separately. The data are available in the Framingham.csv file.

– graphically represent the variations between blood pressure and BMI (bmi) in men and women;

– are the linear correlation coefficients estimated for men and women significantly different at 5%?

– estimate the parameters of the linear regression model considering the blood pressure as a response variable and the BMI as an explanatory variable, for these two subsamples. Give a 95% confidence interval for the estimation of the respective slopes.

Since the data are available in "conventional" CSV (using the comma as field separator), they can be very easily imported by entering the insheet command:

```
. insheet using "Framingham.csv"
. describe, simple
. list in 1/5
```

To facilitate the interpretation and reading of the tables of results, more informative labels will be associated with the categorical variable sex:

```
. label define labsex 1 "M" 2 "F"
. label values sex labsex
. tabulate sex
```

To display a summary of the number of missing data for each of the variables in this data table, we can use the following command:

```
. misstable summarize
```

It can thus be seen that the scl and bmi variables include 33 and 9 missing values, respectively. Hence, the corrected table of the counts by sex:

```
. tabulate sex if bmi < .
```

In Stata, it is not possible to display transparent markers in a scatterplot. In the case where the number of observations is high, it is therefore best to change the default symbol type (filled circle) and display small circles (oh or Oh):

```
. scatter sbp bmi, by(sex) msymbol(Oh)
```

The linear correlation coefficient between the variables sbp and bmi according to sex can be obtained by means of correlate after stratification upon the factor sex. It should be noted that the option by appears in first place and that it is necessary to sort the data as a first step, and the addition of the command sort immediately after stratification:

```
. by sex, sort: correlate sbp bmi
```

For the regression model, we will first transform the bmi (explanatory variable) and sbp (response variable) variables by making use of a logarithmic transformation:

```
. gen logbmi = log(bmi)
. gen logsbp = log(sbp)
```

It is then possible to visually verify through a histogram (or a QQ-plot) that this transformation has been able to correctly reduce the distributions of these two variables close to the normal. In order to simultaneously display the four histograms, each of the figures can be saved in the Stata format (gph) and then be combined with the graph combine command:

```
. histogram bmi, saving(gphbmi)
. histogram logbmi, saving(gphlogbmi)
. histogram sbp, saving(gphsbp)
. histogram logsbp, saving(gphlogsbp)
. graph combine gphbmi.gph gphlogbmi.gph gphsbp.gph gphlogsbp.gph
```

The regression model stratified by gender poses no major problem, and unlike R, it is not necessary to calculate the confidence intervals for the slopes with a separated command because they are directly provided in the table of results returned by Stata:

```
. regress logsbp logbmi if sex == 1, noheader
. regress logsbp logbmi if sex == 2, noheader
```

3) The data available in the quetelet.csv file provide information on systolic blood pressure (PAS), the Quetelet index (QTT), age (AGE) and tobacco consumption (TAB = 1 if smoking, 0 otherwise) for a sample of 32 men over the age of 40.

  – indicate the value of the linear correlation coefficient between systolic blood pressure and the Quetelet index, with a 90% confidence interval;

  – give estimates of the parameters of the regression line of the systolic blood pressure on the Quetelet index;

– test if the slope of the regression line is different from 0 (considering a 5% Type I error);

– graphically represent the variations in blood pressure with respect to the Quetelet index, distinctly showing smokers and non-smokers with symbols or different colors, and draw the regression line whose parameters have been estimated in the preceding regression model;

– repeat the previous analysis by restricting the sample to smokers.

Loading the data raises no specific difficulties because these have been exported from Excel and are in CSV format. However, care should be taken to ensure that the field delimiter type be specified, here a semicolon:

```
. insheet using "quetelet.csv", delim(";") clear
. describe
```

After importing, it can be observed that the variable qtt is not recognized as a number as such but has been coded as a string of characters. This is explained by the fact that the fractional part is separated from the integer portion by a comma and not a dot. It is therefore necessary to convert this variable into a number, which can be achieved with the destring command and the dpcomma option:

```
. destring qtt, dpcomma replace
```

Next, the variable tab is recoded into a qualitative variable including more informative labels:

```
. label define ltab 0 "NF" 1 "F"
. label values tab ltab
. list in 1/5
```

Finally, the numerical summary of the variable is obtained with the summarize command:

```
. summarize pas-tab
```

Since the tab variable is internally coded as $0/1$, the average corresponds to the relative frequency of smokers, that is 53%.

The linear correlation coefficient between the pas and qtt variables can be estimated, as well as its confidence interval, with the corrci command. The level(90) option enables the confidence level to be specified:

```
. corrci pas qtt, level(90)
```

The pwcorr command will be utilized whenever it is desirable to test whether the correlation is zero.

The `regress` command makes it possible to perform a linear regression for a response variable (placed first in the list of variables) and one or more explanatory variables. It is used as follows:

```
. regress pas qtt
```

By default, a table of the ANOVA for the regression model and a table of the model coefficients table are obtained, here the y-intercept (70.58) and the slope of the regression line (21.49). The Student's test associated with the slope allows its statistical significance to be evaluated in the light of the data. When the regression coefficients have to be manipulated, it is possible to extract them from the results table as shown here:

```
. matrix b = e(b)
. svmat b
. display "slope = " b1 ", y-intercept = " b2
```

With regard to displaying the scatterplot with both regression lines superimposed, we utilize a combination of `lfit` (to draw the regression line) and `scatter` (to display the observations). The disadvantage is that it is necessary to manually build the legend. Note that it is required that the regression lines be presented over the whole range of the `qtt` variable, hence the `range(2, 5)` option:

```
. twoway lfit pas qtt if tab == 0, range(2 5) lpattern(dot) ||
  scatter pas qtt if tab == 0, msymbol(square) ||
  lfit pas qtt if tab == 1, range(2 5) ||
  scatter pas qtt if tab == 1, msymbol(circle)
  legend(label(1 "") label(2 "NF") label(3 "") label(4 "F"))
```

The regression of `qtt` on `pas` restricting the analysis to the sole observations of the group smoker (`tab == 1`) does not raise concern because as it has been seen in the previous application, it suffices to add an `if tab == 1` option:

```
. regress pas qtt if tab == 1
```

```
      Source |       SS       df       MS              Number of obs =      17
-------------+------------------------------           F(  1,    15) =   19.40
       Model |  2088.16977     1  2088.16977           Prob > F      =  0.0005
    Residual |  1614.30082    15  107.620055           R-squared     =  0.5640
-------------+------------------------------           Adj R-squared =  0.5349
       Total |  3702.47059    16  231.404412           Root MSE      =  10.374

------------------------------------------------------------------------------
         pas |      Coef.   Std. Err.      t    P>|t|     [95% Conf. Interval]
-------------+----------------------------------------------------------------
```

```
    qtt |   20.11804   4.567193     4.40   0.001     10.3833   29.85278
  _cons |   79.25533   15.76837     5.03   0.000    45.64585   112.8648
```
----

4) Based on the birth weight data [HOS 89], the aim is to study the relationship between the babies' weight (considered as a numeric variable, bwt) and two characteristics of the mother: her weight (lwt) and her ethnic origin (race).

– graphically represent the relationship between the babies' weights and the mothers' weights, according to the ethnicity of the mothers;

– estimate the linear regression parameters considering the baby's weights as a response variable and the mothers' weights centered on their average as an explanatory variable. Is the estimated slope significant at the usual 5% threshold?

– estimate the parameters of the linear regression, where this time the explanatory variable is the mothers' ethnicity, the response variable remaining the babies' weight. Compare the significance of the model in its whole with the results obtained from one-way ANOVA (ethnicity);

– what is the weight predicted for a baby whose mother weighs 60 kg?

– give a 95% confidence interval for a pointwise prediction in average.

Importing data in text format would pose no particular difficulties, but in order to simplify the task we will use the data available online. The dataset on birth weight can be imported by typing webuse as follows:

```
. webuse lbw
```

We will verify that the variables are correctly the same as those found in the birthwt file in R.

To represent the relationship between the babies' and the mothers' weight taking their ethnicity into account, a simple scatter plot can be constructed with the following instructions:

```
. sepscatter bwt lwt, separate(race)
```

Here, we are employing a command downloadable from the SSC site (findit sepscatter) to print out multiple scatterplots in the same graphic using a third conditioning variable. This command avoids combining multiple scatter statements, separated by | |.

For the regression model, an auxiliary variable lwtc has been initially generated, representing the mothers' weight centered on their average, the latter being obtained from summarize. Then, it suffices that the response variable and the explanatory variable be indicated to the regress command as follows:

```
. quietly: summarize lwt
. gen lwtc = lwt/r(mean)
. regress bwt lwtc
```

According to the results, a significant relationship between the two variables is found thereof.

Considering the `race` variable as an explanatory variable, it is necessary to prefix it with the operator `i.` so that Stata properly manipulate the `race` variable as a categorical variable:

```
. regress bwt i.race
```

The reference category is the first level of `race` in order for the regression coefficients printed out to correspond to the mean deviations with respect to the average babies' weight in the "white" category. In comparison, an ANOVA model would give the following results (here, there is no need to specify `i.race`):

```
. oneway bwt race
```

It can be verified that the previous command (or alternatively, simply `regress bwt i.race, notable`) properly yields the same result for the test of all of the regression model.

The predicted value for a child's weight whose mother weighs 60 kg can be obtained from the coefficients of the regression line or by utilizing the `margins` command, which is more flexible to use. In the first case, the postestimation values returned by the `regress` command will be employed (accessible via `ereturn list`), that is:

```
. replace lwt = lwt/2.2
. regress bwt lwt
. display _b[_cons] + _b[lwt]*40
```

In the second case, the instruction to be inserted amounts to:

```
. margins, at(lwt=40)
```

A 95% confidence interval is automatically provided by `margins`. The `level(90)` option would provide a 90% confidence interval.

4

Logistic Regression and Epidemiological Analyses

After having reviewed the principal measures of risk in epidemiologic and diagnostic studies, we will focus on modeling a binary variable according to numerical or binary explanatory variables based on the logistic regression model.

4.1. Measures of association in epidemiology

4.1.1. *Prognostic studies and risk measures*

In addition to `tabodds` mentioned in Chapter 2, Stata provides the command `mhodds` for case-control and cross-sectional studies. Here is an example of how to use it with the variables `low` and `smoke` analyzed in section 2.3:

```
. mhodds low smoke

Maximum likelihood estimate of the odds ratio
Comparing smoke==1 vs. smoke==0

       -----------------------------------------------------------------
        Odds Ratio     chi2(1)        P>chi2         [95% Conf. Interval]
       -----------------------------------------------------------------
          2.021944        4.90         0.0269         1.069897    3.821169
       -----------------------------------------------------------------
```

Now we will consider four age groups for the mothers and will perform a Maentel–Haenszel test to obtain an estimate of the odds ratio by controlling the age:

```
. xtile age4 = age, nq(4)
. table low smoke age4
```

```
--------------------------------------------------------------------
         |           4 quantiles of age and smoke
         | ---- 1 ---     ---- 2 ---     ---- 3 ---     ---- 4 ---
   low   |   0     1        0     1        0     1        0     1
---------+----------------------------------------------------------
     0   |  20    16       26    10       15     6       25    12
     1   |   8     7        8    12       10     5        3     6
--------------------------------------------------------------------
```

It is possible to utilize mhodds low smoke age4 to directly obtain the common odds ratio, but by specifying the stratification factor in an option by() Stata provides the estimates per stratum in addition to the common odds ratio estimate:

```
. mhodds low smoke, by(age4)

Maximum likelihood estimate of the odds ratio
Comparing smoke==1 vs. smoke==0
by age4
```

```
----------------------------------------------------------------------
   age4 | Odds Ratio        chi2(1)         P>chi2     [95% Conf. Interval]
--------+-------------------------------------------------------------
      1 |   1.093750           0.02         0.8856      0.32257    3.70863
      2 |   3.900000           5.50         0.0191      1.14267   13.31098
      3 |   1.250000           0.09         0.7630      0.29217    5.34783
      4 |   4.166667           3.48         0.0619      0.81731   21.24180
----------------------------------------------------------------------
```

```
   Mantel-Haenszel estimate controlling for age4
   -------------------------------------------------------------------
   Odds Ratio     chi2(1)        P>chi2      [95% Conf. Interval]
   -------------------------------------------------------------------
    2.138616        5.59         0.0181       1.121338    4.078767
   -------------------------------------------------------------------
```

```
Test of homogeneity of ORs (approx): chi2(3)   =     3.36
                              Pr>chi2   =   0.3399
```

In this case, we are manipulating individual data, but this command will also work with a table of counts. For this, the weighting will be specified by reporting the counting variable inside the `fweight=` option (see `help mhodds` for examples of how to use it).

In terms of visualization of the stratified tabulated data, it is quite possible to easily build a series of bar graphs employing the `catplot` command presented in section 2.2.1. In order to make the reading of the chart easier, it is necessary to add labels to the three variables being manipulated: `low`, `smoke` and `age4`. For the latter, we need to know the bounds of the class intervals that the `xtile` command has used. This information can be obtained from the `_pctile` command in the following manner. Note that the two extreme bounds are not included during the displaying, but based on `summarize age` we can verify the minimum and the maximum values of this variable. Note that the bounds shown below are inclusive:

```
. _pctile age, n(4)
. return list

scalars:
                r(r1) =    19
                r(r2) =    23
                r(r3) =    26
```

From this, a set of labels can be created for the three variables, and the distribution of numbers corresponding to the three-dimensional array can be displayed. The option `percent` will be included with the `catplot` command when it is preferable to display proportions rather than counts (Figure 4.1):

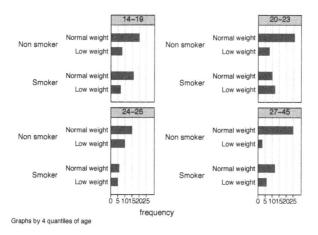

Figure 4.1. *Distribution of children with a smaller weight than the standard according to the mother's age and smoker status*

```
. label define agec 1 "14-19" 2 "20-23" 3 "24-26" 4 "27-45"
. label values age4 agec
. label define wght 0 "Normal weight" 1 "Low weight"
. label values low wght
. label define smoking 0 "Non smoker" 1 "Smoker"
. label values smoke smoking
. catplot low smoke, by(age4)
```

The epitab commands will provide the same result for the calculation of the odds ratio. For example, with the cc command for case-control studies, it would yield:

```
. cc low smoke, by(age4)
```

4 quantiles of an	OR	[95% Conf. Interval]		M-H Weight	
14-19	1.09375	.2719158	4.315057	2.509804	(exact)
20-23	3.9	1.06682	14.50878	1.428571	(exact)
24-26	1.25	.23063	6.531024	1.666667	(exact)
27-45	4.166667	.713997	29.26378	.7826087	(exact)
Crude	2.021944	1.029092	3.965864		(exact)
M-H combined	2.138616	1.130227	4.04669		

```
Test of homogeneity (M-H)      chi2(3) =     3.48  Pr>chi2 = 0.3237

                 Test that combined OR = 1:
                     Mantel-Haenszel chi2(1) =      5.59
                                     Pr>chi2 =      0.0181
```

whereas, if we do not considering the age4 variable:

```
. cc low smoke, woolf
```

	smoke			Proportion	
	Exposed	Unexposed		Total	Exposed
Cases	30	29		59	0.5085
Controls	44	86		130	0.3385
Total	74	115		189	0.3915
	Point estimate			[95% Conf. Interval]	

```
      Odds ratio |      2.021944      |     1.08066    3.783112 (Woolf)
  Attr. frac. ex. |       .5054264      |    .0746392    .7356673 (Woolf)
  Attr. frac. pop |       .2569965      |
                  +---------------------------------------------------
                         chi2(1) =      4.92  Pr>chi2 = 0.0265
```

The response variable is always placed in the first position, followed by the exposure factor. To obtain a measure of the relative risk, cc will be replaced by cs where applicable (cohort study, or even cross-sectional studies).

The following illustrates the low and smoke variables (the risk ratio is first estimated manually using rounded percentages):

```
. tabulate low smoke, col nofreq

                |          smoke
          low |  Non smoke    Smoker |     Total
--------------+----------------------+----------
Normal weight |      74.78     59.46 |     68.78
   Low weight |      25.22     40.54 |     31.22
--------------+----------------------+----------
        Total |     100.00    100.00 |    100.00

. display 40.54/25.22
1.6074544


. cs low smoke

                  |  smoke                 |
                  |  Exposed   Unexposed   |      Total
------------------+------------------------+------------
           Cases |       30          29   |        59
        Noncases |       44          86   |       130
------------------+------------------------+------------
           Total |       74         115   |       189
                  |                        |
            Risk |  .4054054    .2521739   |   .3121693
                  |                        |
                  |   Point estimate       |   [95% Conf. Interval]
                  |------------------------+-----------------------
 Risk difference |        .1532315        |   .0160718     .2903912
      Risk ratio |        1.607642        |   1.057812     2.443262
```

```
Attr. frac. ex. |        .377971        |   .0546528    .5907112
Attr. frac. pop |        .1921887       |
                +---------------------------------------------------
                         chi2(1) =     4.92  Pr>chi2 = 0.0265
```

4.1.2. *Diagnostic studies*

With regard to the evaluation of diagnostic tests, in most cases we will start from a contingency table describing the counts associated with the cross-tabulation of two binary variables. Consider the data originating from a validation study of a new diagnostic test in 1,586 patients. Among the 744 ill patients, 670 have been identified as such by this new test.

The data are reported below. They are available in a Stata file called diagnos.dta and can be directly imported with the use command. Note that in order to avoid losing the current session, preserve is employed to save the data, and then, by removing all (*) variables, the workspace is cleansed. However, it would not be difficult for users to enter themselves the data by means of the built-in editor or the input command:

```
. preserve
. drop *
. use "diagnos.dta"
. list

      +-------------------+
      | Test    Dis    N |
      |-------------------|
  1. |   1      1    670 |
  2. |   0      1     74 |
  3. |   1      0    202 |
  4. |   0      0    640 |
      +-------------------+
```

The frequency table can be reconstructed by weighting the tabulate command:

```
. tabulate Test Dis [fweight=N]

            |         Dis
     Test |        0          1 |      Total
-----------+----------------------+----------
        0 |       640         74 |        714
        1 |       202        670 |        872
```

```
-----------+----------------------+----------
    Total |     842          744 |    1,586
```

From this point, all the information necessary to calculate values such as sensitivity or specificity is available, as well as the positive and negative predictive values. Subsequently, we will see that it is also possible to verify the diagnostic qualities of a test based on a logistic regression model. However, there is a Stata package that automatically calculates all these quantities (use findit diagt and follow the installation procedures).

Here are the results that would be obtained with these data:

```
. diagt Dis Test [fw=N], chi

            |         Test
       Dis |    Pos.        Neg. |     Total
-----------+----------------------+----------
  Abnormal |     670          74 |       744
    Normal |     202         640 |       842
-----------+----------------------+----------
     Total |     872         714 |     1,586

         Pearson chi2(1) = 696.4558   Pr = 0.000
True abnormal diagnosis defined as Dis = 1
```

			[95% Confidence Interval]	
Prevalence	Pr(A)	47%	44%	49.4%
Sensitivity	Pr(+|A)	90.1%	87.7%	92.1%
Specificity	Pr(-|N)	76%	73%	78.9%
ROC area	(Sens. + Spec.)/2	.83	.812	.848
Likelihood ratio (+)	Pr(+|A)/Pr(+|N)	3.75	3.32	4.24
Likelihood ratio (-)	Pr(-|A)/Pr(-|N)	.131	.105	.163
Odds ratio	LR(+)/LR(-)	28.7	21.5	38.2
Positive predictive value	Pr(A|+)	76.8%	73.9%	79.6%
Negative predictive value	Pr(N|-)	89.6%	87.2%	91.8%

Care must be taken not to forget to restore the initial environment by using the appropriate command:

```
. restore
```

4.2. Logistic regression

4.2.1. *Estimation of the model parameters*

When data are available in individual format (long table where each line denotes a statistical unit for which a binary response variable and one or more explanatory variables are accessible), Stata proposes two commands for performing a single or a multiple logistic regression: logit and logistic. They mainly differ in the format of the results they display: by default, logistic returns odds ratios, whereas logit returns the regression coefficients on the log-odds scale.

Note that the probit command allows for the estimation of the parameters of a logistic regression model by considering a link function such as probit rather than logit.

The following gives the result of a logistic regression considering the babies' birth weight indicator (low) as a response variable and the mothers's weight (lwt) as an explanatory variable:

```
. logistic low lwt

Logistic regression                             Number of obs    =        189
                                                LR chi2(1)       =       5.98
                                                Prob > chi2      =     0.0145
Log likelihood = -114.34533                     Pseudo R2        =     0.0255


------------------------------------------------------------------------------
         low |  Odds Ratio   Std. Err.      z    P>|z|     [95% Conf. Interval]
-------------+----------------------------------------------------------------
         lwt |   .9695452    .0131597    -2.28   0.023     .9440927    .995684
       _cons |   2.713702    2.131045     1.27   0.204     .5822659   12.64745
------------------------------------------------------------------------------
```

As in the case of the linear regression, Stata provides the values of the model parameters (here, in terms of the odds ratio associated with the unit variation of the numeric explanatory variable) and their confidence intervals, accompanied by usual significance tests (for the coefficients and the model, LR chi2(1)). The value of pseudo R^2 reported by Stata corresponds to the McFadden coefficient.

Additional goodness of fit indices of the model will be obtained by making use of the fitstat command discussed in Chapter 4:

```
. fitstat

Measures of Fit for logistic of low
```

```
Log-Lik Intercept Only:        -117.336   Log-Lik Full Model:            -114.345
D(187):                         228.691   LR(1):                            5.981
                                          Prob > LR:                        0.014
McFadden's R2:                    0.025   McFadden's Adj R2:                0.008
ML (Cox-Snell) R2:                0.031   Cragg-Uhler(Nagelkerke) R2:       0.044
McKelvey & Zavoina's R2:          0.053   Efron's R2:                       0.032
Variance of y*:                   3.475   Variance of error:                3.290
Count R2:                         0.688   Adj Count R2:                     0.000
AIC:                              1.231   AIC*n:                          232.691
BIC:                           -751.516   BIC':                            -0.740
BIC used by Stata:              239.174   AIC used by Stata:              232.691
```

In the case where the explanatory variable is binary, the modeling principle remains the same. The following illustrates the babies' weight and the presence of intrauterine pain for the mother, this time with the logit command:

```
. logit low ui, nolog

Logistic regression                        Number of obs    =        189
                                           LR chi2(1)       =       5.08
                                           Prob > chi2      =     0.0243
Log likelihood = -114.79795                Pseudo R2        =     0.0216


------------------------------------------------------------------------------
         low |      Coef.   Std. Err.      z    P>|z|     [95% Conf. Interval]
-------------+----------------------------------------------------------------
          ui |   .9469277   .4167734     2.27   0.023     .1300669    1.763789
       _cons |  -.9469277   .1756215    -5.39   0.000     -1.29114   -.6027159
------------------------------------------------------------------------------
```

Note that the option or can be added to obtain the odds ratio, which can be also found by taking the exponential of the regression coefficient stored in _b[ui]:

```
. display exp(_b[ui])
2.5777778
```

4.2.2. Logistic regression and diagnostic studies

In the case of a diagnostic study, or more generally when there is interest in a classification approach rather than a regression one, the following command yields a contingency table summarizing the individuals correctly or incorrectly classified as positive and negative with respect to the response variable, considering a threshold of 0.5 for the detection or events allocation probability:

```
. estat classification

Logistic model for low

                -------- True --------
Classified |        D            ~D  |      Total
-----------+---------------------------+-----------
     +     |       14            14  |        28
     -     |       45           116  |       161
-----------+---------------------------+-----------
   Total   |       59           130  |       189

Classified + if predicted Pr(D) >= .5
True D defined as low != 0
---------------------------------------------------
Sensitivity                    Pr( +| D)    23.73%
Specificity                    Pr( -|~D)    89.23%
Positive predictive value      Pr( D| +)    50.00%
Negative predictive value      Pr(~D| -)    72.05%
---------------------------------------------------
False + rate for true ~D       Pr( +|~D)    10.77%
False - rate for true D        Pr( -| D)    76.27%
False + rate for classified +  Pr(~D| +)    50.00%
False - rate for classified -  Pr( D| -)    27.95%
---------------------------------------------------
Correctly classified                        68.78%
---------------------------------------------------
```

The lroc command allows for displaying the Receiver Operating Characteristic (ROC) curve of the last estimated model as well as the value of the area under the curve.

4.2.3. *Point and interval prediction*

The predict command is used to calculate the values fitted for a given model or to estimate probability or log-odds values for new observations: this is a postestimation command, and it can therefore be utilized after having built a regression model with logit or logistic. The options make it possible to define the type of prediction that we are interested in: when the aim consists of predicting probabilities, the option p will be employed; otherwise, the option xb will provide predictions on the link scale:

```
. quietly: logit low lwt
. predict pr, p
```

The calculation of the 95% confidence intervals for fitted values poses no real difficulty. Manually, we can proceed as follows (on the log-odds scale): as a first step, we generate the linear predictions (`lo`), the prediction error (`lose`) and the bounds of the associated confidence intervals (`lolci` and `louci`). A scatterplot of the predicted values (log odds) according to the mother's weight is then displayed, upon which are superimposed lines delimiting the confidence intervals. It should be observed that it is necessary to order the coordinates of the points in the latter case (Figure 4.2):

```
. predict lo, xb
. predict lose, stdp
. gen lolci = lo - 1.96*lose
. gen louci = lo + 1.96*lose
. twoway (scatter lo lwt, sort connect(l)) (line lolci louci lwt,
          sort pstyle(p3 p3)), ///
    xlabel(35(15)120)
```

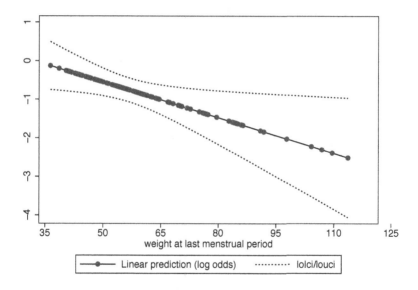

Figure 4.2. *Predicted values for the logistic regression model*

Now consider a model including, in addition to the variable `lwt`, the mothers' ethnicity (`race`). The `margins` command provides very powerful tools to calculate predictions with confidence intervals or marginal effects. Here is a relatively simplified example of how to use it for a model including these two explanatory variables, `lwt` and `race` (Figure 4.3):

```
. logit low lwt i.race
```

```
Iteration 0:    log likelihood =     -117.336
Iteration 1:    log likelihood = -111.73378
Iteration 2:    log likelihood = -111.62959
Iteration 3:    log likelihood = -111.62954
Iteration 4:    log likelihood = -111.62954
```

```
Logistic regression                          Number of obs   =        189
                                             LR chi2(3)      =      11.41
                                             Prob > chi2     =     0.0097
Log likelihood = -111.62954                  Pseudo R2       =     0.0486
```

```
------------------------------------------------------------------------------
         low |      Coef.   Std. Err.      z    P>|z|     [95% Conf. Interval]
-------------+----------------------------------------------------------------
         lwt |  -.0334908   .0141666    -2.36   0.018    -.0612568   -.0057248
             |
        race |
           2 |   1.081066   .4880522     2.22   0.027     .1245015    2.037631
           3 |   .4806032   .3566737     1.35   0.178    -.2184644    1.179671
             |
       _cons |   .8057537   .8451667     0.95   0.340    -.8507426     2.46225
------------------------------------------------------------------------------
```

```
. quietly: margins, at(lwt=(40(10)110)) over(race)
. marginsplot

Variables that uniquely identify margins: lwt race
```

4.2.4. *Case of grouped data*

In the case where the data are grouped, that is when there is only a table available, which summarizes the counts observed for the cross-tabulation of the levels of two categorical variables, we will use the blogit command instead of logit. It is then essential to clarify, in order, the positive events, the total number of events and the associated explanatory variables. Hereafter follows an application example with the same data (low and ui).

This time, instead of using an external file or building the data table with input, we will directly operate on the individual data available in the workspace. All of these manipulations will be supported by a pair of instructions preserve/restore so as to be able to return to the initial data at the end of the example.

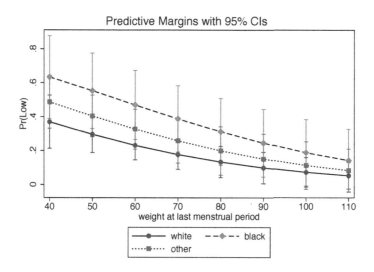

Figure 4.3. *Values predicted with the command* `margins`

Initially, a table is built summarizing the occurrence frequency of each pair of modality for the `low` and `ui` variables by making use of the `contract` command. The `freq(n)` option simply allows the variable used for counting statistics to be renamed. We will then need the total counts associated with each modality of the `ui` variable, which will allow us to obtain the sum of the positive events for the `low` variable (contained in `n`) and the sum of the positive and negative events associated (we call `tot`) when `low=1` ("Low weight"). This can be achieved with the instruction egen, grouped on the modalities of `ui`:

```
. preserve
. contract low ui, freq(n)
. egen tot = sum(n), by(ui)
. list
```

```
      +--------------------------------+
      |            low    ui    n   tot |
      |--------------------------------|
   1. | Normal weight    No  116   161 |
   2. | Normal weight   Yes   14    28 |
   3. |    Low weight    No   45   161 |
   4. |    Low weight   Yes   14    28 |
      +--------------------------------+
```

It is possible to verify that the total counts are correctly identical for each of the modalities of ui, and that the data of interest are the last two rows of this table. This enables for each level of the ui variable that the number of positive events and the number of total events be obtained. The regression model is then formulated as following:

```
. blogit n tot ui if low == 1, or

Logistic regression for grouped data          Number of obs   =        189
                                               LR chi2(1)      =       5.08
                                               Prob > chi2     =     0.0243
Log likelihood = -114.79795                    Pseudo R2       =     0.0216


------------------------------------------------------------------------------
    _outcome | Odds Ratio   Std. Err.      z    P>|z|     [95% Conf. Interval]
-------------+----------------------------------------------------------------
          ui |   2.577778   1.074349     2.27   0.023     1.138905      5.8345
       _cons |    .387931    .068129    -5.39   0.000     .2749573    .5473232
------------------------------------------------------------------------------
```

4.3. Key points

– The regress command is employed in the context of the linear regression, whereas for the logistic regression, logit (or logistic) will be used. In both cases, there is a same group of postestimation commands making it possible to obtain information about the goodness of fit of the model (fitstat). Moreover, they can also calculate fitted values or predict new observations (predict).

– Most measures of risk encountered in epidemiology are accessible with the epitab (cc, cs), or with specific commands (tabodds, mhodds).

4.4. Further reading

As in the case of the linear regression, the book by Vittinghoff et al. [VIT 05] is recommended to deepen the modeling key stages of binary data. For more specific operations on categorical data with Stata, the textbook [HAR 12] is recommended.

4.5. Applications

1) We study the effect of a prophylactic therapy of a macrolide with low doses (Therapy A) on the infectious episodes in patients suffering from cystic fibrosis in a placebo-controlled multicenter randomized trial (B). The results are summarized in Table 4.1:

	Infection		
	No	Yes	Total
Therapy (A)	157	52	209
Placebo (B)	119	103	222
Total	276	155	431

Table 4.1. *Prophylaxis and cystic fibrosis therapy*

– based on a χ^2 test, how can the following question be addressed: can the treatment prevent that infectious episodes occur (considering $\alpha = 0.05$)? Verify that the expected counts are effectively larger than 5;

– can the same conclusion be achieved considering the confidence interval of the odds ratio associated with the treatment effect?

– we aim to verify whether there is a disparity from the point of view of the percentages of infectious episodes according to the center. The data per center are indicated in the table hereafter. Conclude based on a χ^2 test;

	Infection		
	No	Yes	Total
Therapy (A)	51	8	59
Placebo (B)	47	19	66
Total	98	27	125

Center 1

	Infection		
	No	Yes	Total
Therapy (A)	91	35	126
Placebo (B)	61	71	132
Total	152	106	258

Center 2

	Infection		
	No	Yes	Total
Therapy (A)	15	9	24
Placebo (B)	11	13	24
Total	26	22	48

Center 3

– based on the previous table, we aim to verify if the treatment effect is independent of the center or not. We are suggesting to carry out a comparison test between the two treatments controlling for any center effect (Mantel–Haenszel test). Indicate the result of the test as well as the value of the common odds ratio.

First of all, it is necessary to construct the counts table given in the text. A manual input method will serve as a model. On the other hand, we will directly input the labels of the variables and not those of the numeric codes. To do this, it is necessary to indicate to Stata what is the format of these labels by means of the instruction str which is appended with the number of characters that we want to use for the encoding of the modalities of the variables:

```
. clear all
. input str1 treatment str3 infection N

treatment~t  infection  N
1.  "A"  "No"  157
2.  "B"  "No"  119
3.  "A"  "Yes"  52
```

```
4. "B" "Yes" 103
5. end
. list
```

Here follows the statement table, with the associated χ^2 test:

```
. tabulate treatment infection [fweight=N], chi
```

If a more accurate value is desired for the degree of significance of the test, the information returned (in a non-invasive manner) by the previous command can be used:

```
. return list
```

Hence the value of the degree of significance p:

```
. display %10.9f r(p)
```

The above command instructs Stata to display the result in the form of a number with nine decimal places.

Concerning the theoretical counts, the same command as before is employed exactly but adding the option `expected`:

```
. tabulate treatment infection [fweight=N], chi expected nofreq
```

Note that the observed counts are not being displayed by means of the option `nofreq`.

In order to calculate the value of the odds ratio, the command `cc` will be entered. However, the epitab commands require that the variables be encoded in binary format (0 = non-exposed/non-sick, 1 = exposed/sick). The data entry achieved in the previous step being not compatible with this format, it is necessary to recode the data, for example by utilizing the `input` command:

```
. input treatment infection N

     treatm~t  infection N
1. 1 0 157
2. 0 0 119
3. 1 1 52
4. 0 1 103
5. end
. label define tx 0 "Placebo" 1 "Treatment"
. label values treatment tx
. label define yesno 0 "No" 1 "Yes"
. label values infection yesno
```

Then, it is possible to perform the estimate of the odds ratio:

```
. cc infection treatment [fweight=N], woolf exact
```

It should be noted that this time the exact option has been included in order to obtain a Fisher's test instead of the approximation by the χ^2 distribution for the statistical test of the null hypothesis.

When examining the data per center, it is necessary to rebuild the counts tables, only considering the column margins of the count tables given in the text statement. Inputting data with input raises no particular difficulties:

```
. input infection centre N

  infection  centre  N
1. 0 1 98
2. 1 1 27
3. 0 2 152
4. 1 2 106
5. 0 3 26
6. 1 3 22
7. end
. tabulate infection centre [fweight=N], chi
```

Once more we adopt the quick format of aggregated data entry: two variables (infectious episodes yes/no, center number) and the counts associated with the cross-tabulation of each of the levels of these variables. The fweight option then allows for applying a χ^2 test with the tabulate command weighting the two-entry table by the counts N.

To achieve a Mantel–Haenszel test, it is necessary to reconsider all of the data (three 2×2) tables and to make use of the cc command specifying the stratification factor with the option by. What follows is one way to proceed, taking into account three variables: tx (treatment A (1) or B (0)), inf (infection no (0)/yes (1)) and cen (center, 1 to 3). Care should be taken to correctly encode the classes of the two classification variables into 0 and 1:

```
. input tx inf cen N

   tx  inf  cen  N
1. 1 0 1 51
2. 1 1 1 8
3. 0 0 1 47
4. 0 1 1 19
--%<----
13. end
```

It is possible to verify the structure of the data by employing the `table` command, by exactly following the same principle for the options (classification variables, weighting factor and stratification variable). In parallel, we will take the opportunity to add more informative labels to the modalities of the classification variables:

```
. label define txlab 0 "A" 1 "B"
. label define inflab 0 "No" 1 "Yes"
. label values tx txlab
. label values inf inflab
. table tx inf [fw=N], by(cen)
```

With respect to the Mantel–Haenszel test, we obtain the following results:

```
. cc tx inf [freq=N], by(cen)
```

2) Table 4.2 summarizes the proportion of myocardial infarctions observed in men aged 40–59 and for whom the blood pressure and the rate of cholesterol have been measured, considered in the form of ordered classes [EVE 01].

TA	Cholesterol (mg/100 ml)						
	< 200	200 − 209	210 − 219	220 − 244	245 − 259	260 − 284	> 284
< 117	2/53	0/21	0/15	0/20	0/14	1/22	0/11
117 − 126	0/66	2/27	1/25	8/69	0/24	5/22	1/19
127 − 136	2/59	0/34	2/21	2/83	0/33	2/26	4/28
137 − 146	1/65	0/19	0/26	6/81	3/23	2/34	4/23
147 − 156	2/37	0/16	0/6	3/29	2/19	4/16	1/16
157 − 166	1/13	0/10	0/11	1/15	0/11	2/13	4/12
167 − 186	3/21	0/5	0/11	2/27	2/5	6/16	3/14
> 186	1/5	0/1	3/6	1/10	1/7	1/7	1/7

Table 4.2. *Infarction and blood pressure*

The data are available in the `hdis.dat` file in the form of a table comprising four columns that indicate, respectively, the blood pressure (eight categories, rated 1–8), the cholesterol rate (seven categories, rated 1–7), the number of myocardial infarction and the total number of individuals. We are interested in the association between blood pressure and the likelihood of suffering a myocardial infarction:

– calculate the proportions of myocardial infarction for each level of blood pressure and represent them in a table and graphical form;

– based on a logistic regression model, determine whether there is a significant association at $\alpha = 0.05$ between blood pressure, assumed as a quantitative variable considering the class centers, and the likelihood of having a heart attack;

– express in logit units, the probabilities of infarction predicted by the model for each of the blood pressures;

– on the same graph, display the empirical proportions and the logistic regression curve according to the blood pressure values (class centers).

Since the data are presented in the form of a text file in which the fields are separated by spaces, we will enter the `import delimited` command. Caution must be taken to specify that the field separator is composed of one or more spaces, and that the name of the variables appear on the first line of the file:

```
. import delimited "hdis.dat", delimiter(space, collapse) varnames(1)
```

After having verified that the data have been successfully imported, labels can be associated with the numerical codes of the `bpress` variable as follows:

```
. label define bpress 1 "<117" 2 "117-126" 3 "127-136" 4 "137-146" 5 "147-156" 6
        "157-166" 7 "167-186" 8 ">186"
. label values bpress bpress
```

The proportion of events is simply calculated by dividing the `hdis` variable by the `total` variable. Since in this exercise, there is only interest in the systolic pressure, we can immediately aggregate the data as a whole by means of `collapse`. At the same time, it can also be verified that the above calculations are accurate:

```
. collapse (sum) hdis (sum) total, by(bpress)
. generate prop = hdis/total
. list
```

To graphically represent these proportions according to the values of `bpress`, it is possible to make use of a bar or of a dot plot. For example:

```
. graph dot (asis) prop, over(bpress)
```

To convert the `bpress` variable into a numeric variable, we can proceed in the following manner:

```
. recode bpress (1 = 111.5) (2 = 121.5) (3 = 131.5) (4 = 141.5) (5 = 151.5)
        (6 = 161.5) (7 = 171.5) (8 = 181.5)
```

The logistic regression model for grouped data is then written as:

```
. blogit hdis total bpress
```

The `predict` command will then enable us to generate the values predicted by such a model for each observed predictor value (`bpress`). It will be possible to use a dot plot similar to the one utilized above to characterize the empirical proportions in order to represent the probability of myocardial infarction based on blood pressure.

3) A case-control investigation has focused on the relationship between alcohol and tobacco consumption and esophageal cancer in humans (study "Ille and Villaine").

The group of cases consisted of 200 patients suffering from esophagus cancer and diagnosed between January 1972 and April 1974. In total, 775 male controls have been selected from the electoral lists. Table 4.3 shows the distribution of all the subjects according to their daily consumption of alcohol, bearing in mind that a consumption greater than 80 g is considered to be a risk factor [BRE 80]. The data are available in the cc_oesophage.csv file:

	Alcohol intake (g/day)		Total
	≥ 80	< 80	
Cases	96	104	200
Controls	109	666	775
Total	205	770	975

Table 4.3. *Infarction and blood pressure*

– what is the value of the odds ratio and its 95% confidence interval (Woolf method)? Does it provide a good estimate of the relative risk?

– is the proportion of consumers at risk the same among cases and among controls (consider $\alpha = 0.05$)?

– build the logistic regression model that enables testing the association between alcohol consumption and the status of individuals. Is the regression coefficient significant?

– recover the value of the observed odds ratio, in the preceding model, and its confidence interval based on the results of the regression analysis.

The data have been saved in compact format (three columns indicating the presence or not of cancer, the level of alcohol intake and the associated counts):

```
. insheet using "cc_oesophage.csv", clear
. label define yesno 0 "No" 1 "Yes"
. label values cancer yesno
. label define dose 0 "< 80g" 1 ">= 80g"
. label values alcohol dose
. list
```

The previous commands allow loading the data file and that more informative names be associated to the variables modalities (cancer and alcohol). To obtain the total number of patients, the following command can be utilized:

```
. egen ntot = sum(patients)
. display ntot
```

Yet once, this command is not optimal and we will prefer solutions based on the use of scalar and that operate on the results returned by commands such as summarize or tabulate. The ntot variable can be suppressed by typing:

```
. drop ntot
```

The proportion of individuals at risk, which is having a daily consumption of alcohol $\geq$ 80 g is obtained from a simple count table cross-tabulating the variables cancer and alcohol (it is necessary to indicate how the cells have to be filled in by adding the option weight), and requesting the row profiles, that is to say, the relative frequencies per row:

```
. tabulate cancer alcohol [fweight=patients], row
```

The calculation of the odds ratio can be done by means of one of the so-called "immediate" Stata command, or using cc, bearing in mind in this case that the patients column must be taken into account, as previously:

```
. cc cancer alcohol [fweight=patients], woolf
```

If the counts table, that is the distribution of the 975 subjects over the four table cells cross-tabulating the exposure to alcohol and the case-control status, is available it is also possible to make use of the case-control odds ratio calculator accessible by the Stata menus.

To test the hypothesis that the proportion of persons with a daily alcohol intake $\geq$ 80 g is identical in the cases (p_1) and the controls (p_0), the following method can be adopted:

```
. prtesti 96 0.4800 109 0.1406
```

Another way to test this hypothesis consists of noting that the previous hypothesis, $H_0 : \pi_0 = \pi_1$, is true only if the odds ratio is equal to 1, hence the idea of directly performing the χ^2 test for the odds ratio given by the command cc (here, $\chi^2 = 110.26$, $P < 0.001$).

The logistic regression model, similarly to other regression models in Stata, is formulated as follows:

```
. logistic cancer alcohol [freq=patients]
```

By default, Stata presents the results (model coefficients) in the form of odds ratio, with their associated confidence intervals. If it is desirable to directly obtain the regression coefficients (on the log-odds scale), we must have to enter the logit command after using logistic. It would also be possible to directly employ a command of the following type:

```
. logit cancer alcohol [freq=patients]
```

5

Survival Data Analysis

This chapter constitutes an introduction to the modeling of survival data with Stata. After describing the principle of the organization and representation of survival data in Stata, we will focus on the estimation of the survival function with the Kaplan–Meier method and Cox's regression model, which will be then discussed in further detail.

In this chapter, the illustrations are achieved based on a different dataset than the one used so far. The data are available in the R survival package and are exported in CSV format. Missing data are coded as a dot (instead of the value NA, employed by default in R). They can be imported into Stata as follows:

```
. insheet using "lung.csv", clear
(10 vars, 228 obs)
. label define gender 1 "Male" 2 "Female"
. label values sex gender
```

These consist of data on the survival of patients with lung cancer. The variables of interest are the following: time is the survival time in days, status is the status at the point date (1 = censored data, 2 = deceased patient), age is the patient's age and sex is the patient's gender (1 = male, 2 = female). In total, there are 228 patients.

5.1. Data representation and descriptive statistics

5.1.1. *Survival data representation format*

The representation and the modeling of such data rely on a particular class of Stata commands due to the nature of survival data. These are represented by means of a variable encoding a duration and another variable encoding an event of interest (death, relapse, failure, etc.). This namely consists of the sts command that includes several subcommands (`list`, `test`, `graph`, mostly) as well as a few associated commands whose prefix is `st`.

The `stset` command is responsible for creating several auxiliary variables (whose name usually begins with _) that are not directly manipulated but that Stata employs transparently. The same process is used by Stata in the case of data from surveys or from time series data. The `stset` command requires that the variable defining the time durations be indicated (independent of the unit of measure) and with the option `failure()` which represents events encoding. By default, Stata considers that all the non-null and non-missing values signal the event of interest (for instance, death). With 0/1 values, this does not create any problems, assuming that a value of 1 indicates the event.

It can happen, as is the case here, that the death or the event be represented by another value and that censoring is not coded with 0. In this case, it is necessary to specify the value of the death in the option `failure()` as shown below:

```
. stset time, failure(status=2)

    failure event:  status == 2
obs. time interval:  (0, time]
 exit on or before:  failure

-------------------------------------------------------------------------------
   228  total obs.
     0  exclusions
-------------------------------------------------------------------------------
   228  obs. remaining, representing
   165  failures in single record/single failure data
 69593  total analysis time at risk, at risk from t =          0
                            earliest observed entry t =        0
                             last observed exit t =         1022
```

5.2. Descriptive statistics

Here, follows an overview of the raw data, simply obtained with the `list` command:

```
. list time status age sex in 1/5

     +----------------------------+
     | time    status    age    sex |
     |----------------------------|
```

```
1. |   306      2    74    Male |
2. |   455      2    68    Male |
3. | 1010      1    56    Male |
4. |   210      2    57    Male |
5. |   883      2    60    Male |
    +----------------------------+
```

It is naturally possible to make use of the group of Stata commands presented in the previous chapters to achieve the numerical summary of variables. However, care should be taken due to the fact that in this case the data are processed independently of the presence of censoring:

```
. tabulate status, summarize(time)

              |        Summary of time
    status |      Mean    Std. Dev.        Freq.
------------+-----------------------------------
         1 |  363.46032   221.13635          63
         2 |        283   202.80508         165
------------+-----------------------------------
     Total |  305.23246   210.64554         228
```

5.3. Survival function and Kaplan–Meier curve

5.3.1. *Mortality table*

In order to build the mortality table and display the probabilities of survival over time, the `sts list` command will be inserted. Without any other option, Stata displays all temporal events:

```
. sts list
```

However, it is possible to limit the display to certain specific values. To display, for example, the survival probability associated with times 200–300, we could write:

```
. sts list, at(200 300) enter

        failure _d:  status == 2
  analysis time _t:  time

              Beg.                    Survivor     Std.
    Time     Total     Fail          Function    Error     [95% Conf. Int.]
```

```
-------------------------------------------------------------------------
     200        145        72              0.6803   0.0311   0.6149   0.7369
     300         92        29              0.5306   0.0346   0.4605   0.5958
-------------------------------------------------------------------------
```

Note: survivor function is calculated over full data and evaluated at
       indicated times; it is not calculated from aggregates shown at left.

The survival median and its 95% confidence interval can be obtained with the stci command:

```
. stci, dd(2) noshow

          |    no. of
          |  subjects        50%   Std. Err.    [95% Conf. Interval]
----------+--------------------------------------------------------------
    total |       228     310.00       21.77         284         361
```

The noshow option does not allow for displaying reminders concerning the variables being used (here, time and status). Regarding the dd(2) option, it sets the limit of the display to two decimal places. If it is desirable to obtain the 10th percentile rather than the median, p(10) will have to be specified in option:

```
. stci, p(10) dd(2)

        failure _d:  status == 2
  analysis time _t:  time

          |    no. of
          |  subjects        10%   Std. Err.    [95% Conf. Interval]
----------+--------------------------------------------------------------
    total |       228      79.00       14.94          54         105
```

This command can also be employed when there are several groups and when their survival median have to be compared. In this case, the classification variable will be indicated by using the by() option:

```
. stci, by(sex) noshow

          |    no. of
sex       |  subjects        50%   Std. Err.    [95% Conf. Interval]
----------+--------------------------------------------------------------
     Male |       138        270    26.78831         210         306
   Female |        90        426    44.20601         345         524
----------+--------------------------------------------------------------
    total |       228        310    21.77251         284         361
```

5.3.2. *Kaplan–Meier curve*

The `sts graph` command makes it possible to graphically represent the survival curve of one or more samples. In the case of several samples, the classification criterion is indicated by means of the `by()` option. The basic syntax is therefore (Figure 5.1):

```
. sts graph
```

Note that we can simply type `sts`.

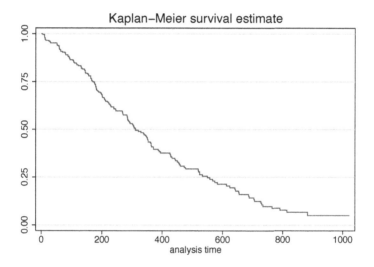

Figure 5.1. *Kaplan–Meier survival curve*

The `ci` option will be added to display confidence intervals (for each time value). Here is an example of its use including other options such as the simultaneous display of the number of individuals at risk over time (by default, Stata employs the same time coordinates than those displayed for the x-axis but this can be modified). Another interesting option is `censored()` (it can be abbreviated into `cen()`) that overlays on the survival curve the observed censored data (Figure 5.2):

```
. sts graph, noshow ci risktable censored(single)
```

In the case of two samples, we will include the classification factor via the option `by()`. The rest of the options is applicable. In Figure 5.3, the position of the caption has been modified such that it appear inside the graph and not in the lower margin of the chart:

```
. sts graph, by(sex) legend(ring(0) position(2))

        failure _d:  status == 2
   analysis time _t:  time
```

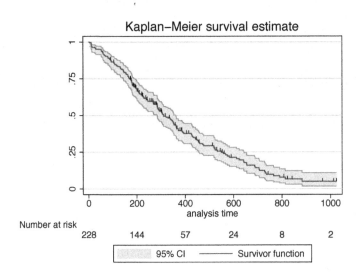

Figure 5.2. *Kaplan–Meier survival curve with confidence intervals and number of individuals at risk*

5.3.3. *Cumulative hazard function*

If we are willing to work with the cumulative hazard function (most often denoted $H(t)$), it suffices to add the cumhaz option when the sts graph command is employed:

```
. sts graph, noshow cumhaz ci
```

5.3.4. *Survival functions equality test*

The sts list command provides the mortality table and the estimated survival values for each time. The by() option makes it possible to calculate the survival function for two or more groups of individuals. However, it is also possible to couple this by() option to the compare option to directly display the survival estimated in each of the groups side by side:

```
. sts list, by(sex) compare noshow
```

		Survivor Function	
sex		Male	Female
time	5	1.0000	0.9889
	132	0.7681	0.8993
	259	0.5192	0.7186
	386	0.3265	0.5089
	513	0.2232	0.4110
	640	0.1228	0.3433
	767	0.0781	0.0832
	894	0.0357	0.0832
	1021	0.0357	.
	1148	.	.

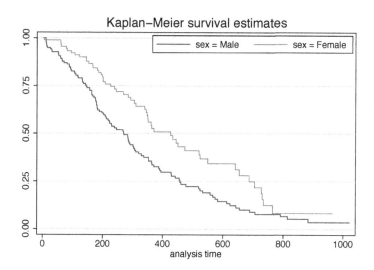

Figure 5.3. *Kaplan–Meier survival curve for two samples*

To perform a log-rank test (equality of the survival functions), we will include the command `sts test` simply indicating the variable defining the groups that have to be compared:

```
. sts test sex, noshow
```

```
Log-rank test for equality of survivor functions
-------------------------------------------------

         |   Events        Events
sex      |  observed      expected
---------+-------------------------
Male     |      112         91.58
Female   |       53         73.42
---------+-------------------------
Total    |      165        165.00

           chi2(1)  =      10.33
           Pr>chi2  =     0.0013
```

If we would rather perform a Wilcoxon test, the option `wilcoxon` will be added as follows:

```
. sts test sex, wilcoxon noshow

Wilcoxon (Breslow) test for equality of survivor functions
-----------------------------------------------------------

         |   Events        Events       Sum of
sex      |  observed      expected       ranks
---------+-----------------------------------------
Male     |      112         91.58        3148
Female   |       53         73.42       -3148
---------+-----------------------------------------
Total    |      165        165.00          0

           chi2(1)  =      12.47
           Pr>chi2  =     0.0004
```

5.4. Cox regression

The command to perform a Cox regression is `stcox`. Its use is substantially identical to that of the regression commands seen in previous chapters, except that it is not necessary to specify a response variable: as for the other `sts` commands, Stata transparently manages the time/event representation. It will therefore suffice to indicate the list of explanatory variables after the name of the command. Here is an example of how to use it considering the variable `sex` only:

```
. stcox sex, noshow
```

```
Iteration 0:   log likelihood = -750.12202
Iteration 1:   log likelihood = -744.83027
Iteration 2:   log likelihood = -744.81818
Iteration 3:   log likelihood = -744.81818
Refining estimates:
Iteration 0:   log likelihood = -744.81818


Cox regression -- Breslow method for ties

No. of subjects =           228              Number of obs   =        228
No. of failures =           165
Time at risk    =         69593
                                             LR chi2(1)      =      10.61
Log likelihood  =   -744.81818              Prob > chi2      =     0.0011


-------------------------------------------------------------------------------
         _t | Haz. Ratio   Std. Err.      z    P>|z|     [95% Conf. Interval]
------------+------------------------------------------------------------------
        sex |   .5883716   .0983645    -3.17   0.002     .4239817    .8165002
-------------------------------------------------------------------------------
```

To indicate the presence of a stratification factor, the strata() option will be entered as follows:

```
. stcox age, strata(sex) noshow

Iteration 0:   log likelihood = -643.61669
Iteration 1:   log likelihood = -642.03076
Iteration 2:   log likelihood = -642.02946
Refining estimates:
Iteration 0:   log likelihood = -642.02946


Stratified Cox regr. -- Breslow method for ties

No. of subjects =           228              Number of obs   =        228
No. of failures =           165
Time at risk    =         69593
                                             LR chi2(1)      =       3.17
Log likelihood  =   -642.02946              Prob > chi2      =     0.0748


-------------------------------------------------------------------------------
         _t | Haz. Ratio   Std. Err.      z    P>|z|     [95% Conf. Interval]
```

```
-------------+--------------------------------------------------------------
        age |   1.016324    .0093351    1.76   0.078     .998191   1.034786
-------------+--------------------------------------------------------------
                                                            Stratified by sex
```

Note that it is possible to change how Stata addresses tied observations, by specifying the efron option for example.

If we want to display the regression coefficients rather than the hazard ratio, the nohr option ought to be specified:

```
. stcox sex, noshow nolog nohr

Cox regression -- Breslow method for ties

No. of subjects =         228                Number of obs   =        228
No. of failures =         165
Time at risk    =       69593
                                             LR chi2(1)      =      10.61
Log likelihood  =  -744.81818                Prob > chi2     =     0.0011


-------------+--------------------------------------------------------------
         _t |      Coef.   Std. Err.      z    P>|z|     [95% Conf. Interval]
-------------+--------------------------------------------------------------
        sex |  -.5303966    .1671808   -3.17   0.002    -.858065   -.2027282
-------------+--------------------------------------------------------------
```

The noshow and nolog options are used to suppress the display of survival variables and the iterations for the convergence of the model.

5.5. Key points

– The representation of censored data is carried out via the command stset that allows the variables encoding times and events to be defined.

– A set of subcommands is associated with sts and other commands are prefixed by st (for example stci).

– The construction of the Cox regression model follows the same principle as in the case of linear or logistic regression, and postestimation commands enable additional information to be provided (predicted values, goodness of fit of the model, etc.).

5.6. Further reading

The book by Cleves *et al.* [CLE 10] definitely remains the reference book for processing survival data with Stata. For a more in-depth coverage of survival analysis, see Royston and Lambert's work [ROY 11].

5.7. Applications

1) In a placebo-controlled trial on biliary cirrhosis, D-penicillamine (DPCA) has been introduced in the active arm in a cohort of 312 patients. In total, 154 patients have been randomized in the active arm (variable treatment, rx, 1 = Placebo, 2 = DPCA). A data set comprising age, biological data and varied clinical signs including the level of bilirubin serum (bilirub) are available in the pbc.txt file [VIT 05]. The patient's status is stored in the variable status (0 = alive, 1 = deceased) and the follow-up duration (years) represents the elapsed time in years since the date of the diagnosis.

– How many deceased individuals can be identified? What proportion of these deaths can be found in the active arm?

– display the distribution of the follow-up durations of 312 patients, by distinctively bringing forward the deceased individuals. Calculate the median follow-up time (in years) for each of the two treatment groups. How many positive events are there beyond 10.5 years and what is the gender of these patients?

– the 19 patients, whose number (number) appears among the following list have undergone a transplant during the follow-up period:

```
5 105 111 120 125 158 183 241 246 247 254 263 264 265 274 288 291
295 297
```

Indicate their average age, the distribution according to sex and the median duration of the follow-up in days until transplant.

– display a table summarizing the distribution of individuals at risk according to time, with the associated survival value;

– display the Kaplan–Meier curve with a 95% confidence interval, without considering the treatment type;

– calculate the survival median and its 95% confidence interval for each group of subjects and display the corresponding survival curves;

– perform a log-rank test considering as predictor the factor rx. Compare with a Wilcoxon test;

– carry out a log-rank test on the factor of interest (rx) by stratifying the age. Three age groups will be considered: 40 years old or less, between 40 and 55 years of age inclusive, more than 55 years old;

– try to find the results of exercise 1(g) with a Cox regression.

The data file is a text file with tabs as field separator. It can be imported in Stata utilizing the insheet command. To display the name of the variables after importing it, it suffices to enter describe with the option simple:

```
. insheet using "pbc.txt", tab
. describe, simple
```

After recoding the labels of the rx and sex variables:

```
. label define trt 1 "Placebo" 2 "DPCA"
. label define sexe 0 "M" 1 "F"
. label values rx trt
. label values sex sexe
```

the proportion of patients who died can be verified (status, 0 = alive and 1 = deceased) and their distribution according to the treatment group based on simple and crossed tabulation. Regarding the cross-tabulation, the option row will be added to obtain the relative frequencies per status:

```
. tabulate status
. tabulate status rx, row
```

To display the distribution of the follow-up times, we will make use of a simple scatterplot. To distinctly present the observations according to the status (0 or 1), we could very well superimpose two sets of points over the same graph. Here is another way of proceeding:

```
. separate number, by(status)
. twoway scatter number0 number1 years, msymbol(S O)
```

The first command actually allows the numbers of patients to be separated according to the status function in order to display both sets of observation with respect to the follow-up durations in years.

The median of the follow-up duration per treatment group can be obtained with the tabstat command operating on a group basis with the option by:

```
. tabstat years, by(rx) stats(median) nototal
```

The number of deaths beyond 10.5 years of follow-up is obtained with a simple tabulation inserting the tabulate command:

```
. tabulate status if years > 10.49
```

as well as the gender of patients who died after this period:

```
. tabulate sex if years > 10.49 & status == 1
```

Regarding the analysis of transplant patients, it is possible to restrict the data table to these patients only. Since Stata allows working with only one data table at the time, it is however necessary to temporarily save the current data before creating a new table:

```
. preserve
. egen idx = anymatch(number), values(5 105 111 120 125 158 183 241 246 247 254
  263 264 265 274 288 291 295 297)
. keep if idx
. gen days = years*365
. tabstat age sex days, stats(mean median sum)
```

The first command makes it possible to build a list of individuals that we may want to employ to filter the original data table (based on subject IDs contained in the variable number). Next, the calculation of the descriptive statistics is carried out utilizing the command tabstat. Once the calculations have been completed, the original data can be restored in the following manner:

```
. restore
```

Stata makes use of its own conventions for encoding survival data. The essential commands are thus: stset to define the way in which events are logged and the observation time, sts to calculate a survival table based on the Kaplan–Meier estimator. Here is how to apply these commands to build the table and the survival curve, regardless of the treatment factor:

```
. stset years, failure(status)
. sts list
```

The second command displays the requested table. For the survival curve, we will use:

```
. sts graph, ci censored(single)
```

The option ci enables displaying the 95% confidence interval for the Kaplan–Meier estimator.

The survival median is obtained by employing the command stci. Without further information, Stata calculates the median survival and its 95% confidence interval for all of the observations. The by() option will be added to obtain the median survival per treatment group:

```
. stci, by(rx)
```

Similarly, it is possible to print the two corresponding survival curves with `sts graph` retaining the `by()` option:

```
. sts graph, by(rx) cen(single)
```

With regard to the log-rank test, the command `sts test` has to be entered, the `logrank` option being the default option. We simply need to indicate the classification variable to test the equality of the survival functions, that is:

```
. sts test rx
```

The `wilcoxon` option will be added to obtain the Wilcoxon test. The `noshow` option allows simplifying the text output by removing the information relative to the variables `sts`:

```
. sts test rx, wilcoxon noshow
```

The same command can be used when we want to include a stratification factor, with the `strata()` option. As a first step, the variable `age` is recoded into a four-class qualitative variable using `egen`:

```
. egen agec = cut(age), at(26,40,55,79)
. sts test rx, strata(agec) noshow
```

Finally, to achieve a Cox regression model, with stratification on the `agec` factor, we employ the command:

```
. stcox rx, strata(agec)
```

By default, the results returned by Stata are expressed in terms of risk (`Haz. Ratio`). In order to obtain the regression coefficients, it is necessary to add the `nohr` option:

```
. stcox rx, strata(agec) nohr
```

2) In a randomized clinical trial, the aim was to compare two treatments for prostate cancer. Patients took orally each day either 1 mg of diethylstilbestrol (DES, active arm) or a placebo, and the survival time is measured in months [COL 94]. The question of interest is knowing whether the survival is different between the two groups of patients, and the other variables present in the `prostate.dat` data file will be ignored.

– Calculate the survival median for the totality of the patients and per treatment group.

– what is the difference between the survival proportions in both groups at 50 months?

– display the survival curves for the two groups of patients;

– perform a log-rank test to verify the hypothesis according to which the DES treatment has a positive effect on the survival of patients.

The text format of the data file prostate.dat is identical to that of the file pbc.txt of the previous exercise, except that the fields are separated by a single space. The command insheet will thus be used:

```
. insheet using "prostate.dat", delimiter(" ")
. list in 1/5
```

The number of living patients at the point time is obtained from tabulate:

```
. tabulate status
```

that is six persons still alive at the end of follow-up time.

To tell Stata which variables are utilized to identify the events (status) and the time (time), we use the command stset in the following manner:

```
. stset time, failure(status)
```

The median survival according to the treatment is available with the command stci specifying the percentile of interest (here, p(50)):

```
. stci, by(treatment) p(50)
```

To plot the survival curves for each treatment arm, we will make use of the command sts graph specifying the classification factor by means of the option by. The option censored displays the censored data:

```
. sts graph, by(treatment) censored(s)
```

The log-rank test is achieved with the command sts test. Note that only the treatment factor has to be specified, the notion of response variable being managed from the beginning by stset:

```
. sts test treatment, noshow
```

Regarding Cox's model, the stcox command will be employed as in the previous exercise, bearing in mind that by default Stata provides the estimated value for the risk, and not the regression coefficient (use nohr):

```
. stcox treatment, noshow
```

Bibliography

[ACO 14] ACOCK A., *A Gentle Introduction to Stata*, 4th ed., Stata Press, College Station, TX, 2014.

[BAU 16] BAUM C., *An Introduction to Stata Programming*, 2nd ed., Stata Press, College Station, TX, 2016.

[BLI 52] BLISS C., *The Statistics of Bioassay*, Academic Press, New York, 1952.

[BRE 80] BRESLOW N., DAY N., *Statistical Methods in Cancer Research: Vol. 1, The Analysis of Case-Control Studies*, IARC Scientific Publications, Lyon, 1980.

[CLE 10] CLEVES M., GOULD W., GUTIERREZ R. *et al.*, *An Introduction to Survival Analysis Using Stata*, 3rd ed., Stata Press, College Station, TX, 2010.

[COL 94] COLLETT D., *Modelling Survival Data in Medical Research*, Chapman & Hall/CRC, Boca Raton, 1994.

[DUP 09] DUPONT W., *Statistical Modeling for Biomedical Researchers*, 2nd ed., Cambridge University Press, Cambridge, 2009.

[EVE 01] EVERITT B., RABE-HESKETH S., *Analyzing Medical Data using S-PLUS*, Springer, New York, 2001.

[FRY 14] FRYDENBERG J., *An Introduction to Stata for Health Researchers*, 4th ed., Stata Press, College Station, TX, 2014.

[HAM 13] HAMILTON L., *Statistics with Stata: Version 12*, 8th ed., Cengage, Belmont, CA, 2013.

[HAN 93] HAND D., DALY F., MCCONWAY K. *et al.* (eds), *A Handbook of Small Data Sets*, Chapman & Hall/CRC, Boca Raton, 1993.

[HAR 12] HARDIN J., HILBE J., *Generalized Linear Models and Extensions*, 3rd ed., Stata Press, College Station, TX, 2012.

[HOS 89] HOSMER D., LEMESHOW S., *Applied Logistic Regression*, John Wiley & Sons, New York, 1989.

[LAL 16] LALANNE C., MESBAH M., *Biostatistics and Computer-based Analysis of Health Data using R*, ISTE Press Ltd, London and Elsevier Ltd, Oxford, 2016.

[MIT 12] MITCHELL M., *Interpreting and Visualizing Regression Models Using Stata*, Stata Press, College Station, TX, 2012.

[PEA 05] PEAT J., BARTON B., *Medical Statistics: A Guide to Data Analysis and Critical Appraisal*, 2nd ed., John Wiley & Sons, New York, 2005.

[RAB 04] RABE-HESKETH S., EVERITT B., *A Handbook of Statistical Analyses using Stata*, 3rd ed., Chapman & Hall/CRC, Boca Raton, 2004.

[ROY 11] ROYSTON P., LAMBERT P., *Flexible Parametric Survival Analysis using Stata: Beyond the Cox Model*, Stata Press, College Station, TX, 2011.

[SEL 98] SELVIN S., *Modern Applied Biostatistical Methods using S-PLUS*, Oxford University Press, New York, 1998.

[STU 08] STUDENT, "The probable error of a mean", *Biometrika*, vol. 6, no. 1, pp. 1–25, 1908.

[VIT 05] VITTINGHOFF E., GLIDDEN D., SHIBOSKI S. *et al*, *Regression Methods in Biostatistics. Linear, Logistic, Survival, and Repeated Measures Models*, Springer, New York, 2005.

Index

Printed in the United States
By Bookmasters